Molecular Modeling

by Hans-Dieter Höltje and Gerd Folkers

This publication has been generously supported by Tripos GmbH, Munich.

Methods and Principles in Medicinal Chemistry

Edited by
R. Mannhold
H. Kubinyi
H. Timmerman

Editorial Board

F. Darvas, T. Fujita, C. R. Ganellin,
F. Gualtieri, U. Hacksell, H.-D. Höltje,
G. Leclerc, R. Rekker, J.-K. Seydel,
D. Triggle, H. van de Waterbeemd

Molecular Modeling

Basic Principles and Applications

by Hans-Dieter Höltje and Gerd Folkers

in collaboration with Thomas Beier,
Wolfgang Sippl and Didier Rognan

To Yvonne Martin

San Francisco, 13.9.97

Gerd Folkers

Hans-Dieter Höltje

VCH Weinheim · New York · Basel · Cambridge · Tokyo

Series Editors:

Prof. Dr. Raimund Mannhold
Biomedical Research Center
Molecular Drug Research Group
Heinrich-Heine-Universität
Universitätsstraße 1
D-40225 Düsseldorf
Germany

Prof. Dr. Hugo Kubinyi
ZHV/W, A 30
BASF AG
D-67056 Ludwigshafen
Germany

Prof. Dr. Hendrik Timmerman
Faculty of Chemistry
Dept. of Pharmacochemistry
Free University of Amsterdam
De Boelelaan 1083
NL-1081 HV Amsterdam
The Netherlands

Authors:

Prof. Dr. H.-D. Höltje
Institute of Pharmaceutical
Chemistry
Heinrich-Heine-Universität
Universitätsstraße 1
D-40225 Düsseldorf
Germany

Prof. Dr. G. Folkers
Department of Pharmacy
ETH Zürich
Winterthurer Str. 190
CH-8057 Zürich
Switzerland

The financial support of Tripos GmbH, Munich, is gratefully acknowledged.

Published jointly by
VCH Verlagsgesellschaft mbH, Weinheim (Federal Republic of Germany)
VCH Publishers, Inc., New York NY (USA)
Editorial Director: Dr. Michael Bär

Library of Congress Card No. applied for.
British Library Cataloguing-in-Publication Data: A catalogue record for this book is available from the British Library.
Deutsche Bibliothek Cataloguing-in-Publication Data:
Höltje, Hans-Dieter:
Molecular modeling : basic principles and applications / by Hans-Dieter Höltje and Gerd Folkerts. In collab. with Thomas Beier ... – Weinheim ; New York ; Basel ; Cambridge ; Tokyo : VCH, 1996
  (Methods and principles in medicinal chemistry ; Bd. 5)
  ISBN 3-527-29384-1
NE: Folkerts, Gerd:; GT

Composition: Kühn & Weyh, D-79111 Freiburg
Printing: Betz Druck GmbH, D-64291 Darmstadt
Bookbinding: Großbuchbinderei J. Schäffer GmbH & Co KG, D-67269 Grünstadt
Printed in the Federal Republic of Germany.

Distribution:
VCH, P.O. Box 10 11 61, D-69451 Weinheim (Federal Republic of Germany)
Switzerland: VCH, P.O. Box, Ch-4020 Basel (Switzerland)
United Kingdom und Irland: VCH (UK) Ltd., 8 Wellington Court, Cambridge CB1 1HZ (England)
USA and Canada: VCH, 220 East 23rd Street, New York, NY 10010–4606 (USA)
Japan: VCH, Eikow Building, 10-9 Hongo 1-chome, Bunkyo-ku, Tokyo 113 (Japan)

Preface

The fifth volume of the series "Methods and Principles in Medicinal Chemistry" focuses on molecular modeling. Progress in modern ligand design is intimately coupled with the access to and the continuous refinement of molecular modeling techniques. They allow the computer-aided generation of molecular structures as well as the computation of molecular properties. Predictions of the three-dimensional structures of drug and receptor molecules, visualizations of their molecular surface properties and optimizations of drug–receptor interactions by visual inspection can be realized today.

The present volume offers an introduction to the field of molecular modeling. The book is organized in two parts: the first deals with the modeling of small molecules whereas the second examines biological macromolecules, in particular proteins.

The first part describes in detail the basic know-how necessary for generating 3D coordinates of small molecules, the computational tools for geometry optimization and conformational analysis, the determination of molecular interaction potentials, approaches for the identification of pharmacophores and last but not least the use of databases. The application of this spectrum of methodical approaches is exemplified by a case study dealing with pharmacophore definition in the field of serotonin receptor ($5 HT_{2A}$) ligands.

The second part gives an introduction to protein modeling. After a description of terminology and the principles governing protein structures, approaches for knowledge-based protein modeling are summarized, followed by chapters on refinement and validation of protein models and on methods for the description of structural properties of proteins. The case study in the second part illustrates the application of experimental procedures to the modeling of protein–ligand complexes (design of non-natural peptides as high-affinity ligands for a MCH I protein).

The editors would like to thank the contributors for their encouragement in compiling this volume. We are sure that scientists entering the fascinating field of computer-aided ligand design will find in this volume the adequate support they need to apply molecular modeling techniques successfully.

April 1996

Düsseldorf   Raimund Mannhold
Ludwigshafen   Hugo Kubinyi
Amsterdam   Hendrik Timmerman

Methods and Principles in Medicinal Chemistry

Edited by
R. Mannhold
H. Kubinyi
H. Timmerman

A Personal Foreword

"A Model must be wrong, in some respects, else it would be the thing itself. The trick is to see where it is right."

Henry A. Bent

We humans receive our data through the senses of vision, touch, smell, hearing and taste. Therefore, when we have to understand things that happen on the submicrosopic scale, we have to devise a way of simulating this activity. The most immediate and accessible way to represent the world that is unobservable is to make a model that is on our scale and that uses familiar forms.

Many physical and chemical properties and behaviors of molecules can be predicted and understood only if the molecular and electronic structures of these species are conceived and manipulated in three-dimensional (3D) models. As a natural follow up nowadays the computer is used as a standard tool for generating molecular models in many research areas.

The historical process of developing concepts leading to molecular modeling started with the quantum chemical description of molecules. This approach yields excellent results on the ab initio level. But the size of the molecular systems which can be handled is still rather limited. It is therefore that the introduction of molecular modeling as a routine tool owes its beginning to the development of molecular mechanics some 25 years ago together with the appearance of new technologies in computer graphics.

The goal of this book is to show how theoretical calculations and 3D visualization and manipulation can be used not simply to look at molecules and take pretty pictures of them, but actually to be able to gain new ideas and reliable working hypotheses for molecular interactions such as drug action.

It is our intention to reach this goal by giving examples from our own research fields more than reporting literature's success stories. This is because stepwise procedures avoiding pitfalls and overinterpretation can at best be demonstrated by data from our own laboratory notebooks.

Most of the contents will therefore reflect our own ideas and personal experiences, but nevertheless represent, what we believe to be an independent view of molecular modeling.

We gratefully acknowledge the technical assistance of Matthias Worch, Frank Alber and Oliver Kuonen. Finally we wish to express our sincere gratitude to Heide Westhusen for her excellent secretarial and organizational help.

Spring 1996

Berlin   Hans-Dieter Höltje
Zürich   Gerd Folkers

Contents

1 Introduction

"Dear Venus that beneath the gliding stars ..." Lukrez (Titus Lucretius Carus, 55 B.C.) starts his most famous poem *De Rerum Natura* with the wish to the Goddess of love to reconcile the wargod Mars, which in this time when the Roman Empire starts to pass over its zenith, ruled the world.

Explanation is the vision of Lukrez. His aim is in odd opposition to his introductory wish to the goddess of love: the liberation of people from his fear of God, from the dark power of unbelievable nature.

The explanation of mechanism from the common is the measure with which Lukrez will take away the fear from the ancient people, the fear of the gods and their priests, the fear of the want of nature and the power of the stars.

Lightning, fire and light, wine and olive oil have been perhaps the simple things of daily experience, which people needed, which people was afraid of, whom has been dear to him:

> "... again, light passes through the horn
> of the lantern's side, while rain is dashed away.
> And why? – unless those bodies of light should be
> finer than those of water's genial showers.
> We see how quickly through a colander
> the wines will flow; how, on the other hand
> the sluggish olive oil delays: no doubt
> because 'tis wrought of elements more large,
> or else more crook'd and intertangled ..."

The atom theory of Demokrit leads Lukrez to the description of the quality of light, water and wine. For this derivation of structure–quality relationships he uses models. The fundamental building stones of Lukretian models look a little like our atoms, called *primodials* by Lukrez, elementary individuals, which were not cleavable anymore. Those elementary building stones could associate. Lukrez even presupposes recognition and interaction. He provides his building stones with mechanic tools that guarantee recognition and interaction. The most important of these conceptual tools are the complementary structure (sic!) and the barked hook. With these primordials Lukrez built his world.

How well the modeling fits is shown in his explanation of the fluidity of wine and oil. A comparison of the space-filling models of the fatty acid and water molecules amazes, because of its similarity with the 2000-years old image of Lukrez.

1.1 Modern History of Molecular Modeling

The roots from which the methods of modern molecular modeling have developed, lie at the beginning of our century, the first successful representations of molecular structures being closely linked to the rapid developments in nuclear physics.

Crystallography was the decisive line of development of molecular modeling. Knowledge of the complexity of crystal structures increased very rapidly but their solution still required huge arithmetic expense to produce only an inadequate two-dimensional (2D) paper representation. The use of molecular kits was the only possible way of obtaining a 3D impression of crystal structure.

The Dreiding Models became famous because they contained all the knowledge of structure chemistry at the time. Prefabricated modular elements, for example different nitrogen atoms with the correct number of bonds and angles corresponding to their hybridization state, or aromatic moieties, made it possible to build up very exact 3D models of the crystal structures, thus allowing molecular modeling. Dimensions were translated linearly from the Ångstrom area. Steric hindrances of substituents, hydrogen bond interactions, etc. were quite well represented by the models. A similar quality of modeling, albeit less accurate—but space filling—was provided by Stuart–Briegleb or CPK models. Watson and Crick described their fumbling with such molecular kits and self-constructed building parts, first to model base pairing and eventually, to outline the DNA helix.

Molecular modeling is not a computer science a priori, but does the computer provide an additional dimension in molecular modeling/molecular design? Indeed, development of the computer occured synergetically, as faster and faster processors repeated the necessary computational steps in shorter and shorter times so that proteins containing thousands of atoms can easily be handled today. However, the molecular graphics technology looked for a further quantum leap bound to the same fast processors. For the first time, in the 1970s the pseudo 3D description of a molecule, color-coded and rotatable, was possible on the computer screen. "Virtual Dreiding models" had been created. Without computer technology the flood of data emerging from a complex structure such as a protein would have exceeded the saturation limits of human efficiency. Proteins would not have been measurable with methods such as X-ray structure analysis and nuclear magnetic resonance without the corresponding computer technology. Indeed, it is computer technology that has made these methods what they are today.

There is however a second factor, without which today's computer-assisted molecular design would be unthinkable. Since the 1930s, nuclear physics has required not only analytical but also systematic thought, a component that was vital in construction of the atomic bomb. Consequently, mathematical modeling techniques were employed for the computation of physical states, and even their prediction.

In the 1940s the computers in Los Alamos were, in the true sense of the word, made of soldiers. Gathered in large groups, everyone had to solve a certain calculation step, but always the same step for the same man. It was here that computer development sought a revolution. The Monte Carlo Simulation, which originated at that time, was applied to the prediction of physical states of gas particles. From that time also the first applications of mechanical analogies on molecular systems were developed. The force fields were born and optimized and, in the course of time have achieved the unbelievable efficiency of modern times.

Mathematical approximation techniques have now made possible the quantum chemical calculation of systems even larger than the hydrogen atom, permitting "quantum dynamic" simulations of ligand binding at the active site of enzymes.

1.2 Do Today's Molecular Modeling Methods Illustrate only the Lukretian World?

This is in fact a question of quality of use. The methods could be used naively or intelligently, though the results are clearly distinguishable. However, naive uses should not be condemned, as it is vital for the quality of the use that a sufficiently critical position is taken when examining the results. In other words, the user realizes his or her naive use of the methods. Now, the researcher is conscious of the restrictions of the method and knows how to judge the results. Here, even with a very simple approach, this critical position results in further knowledge of the correlation between structure and properties.

Often however, such a critical attitude is not present—perhaps the result of modern commercial modeling systems. Those programs always provide a result, the evaluation of which is at liberty of the user. The programs tend stubbornly to calculate every absurd application and present a result—not only a number, but also a graph—and represent a further instrument of seduction for the uncritical use of algorithms. In contrast, the merits of molecular graphics is undisputed because of their essential contribution to the development of other analytical methods such as nuclear magnetic resonance spectroscopy and the X-ray analysis of proteins.

The tendency to perfect data presentation is the reverse situation. For example, visualization of isoelectrical potentials is one of the most valuable means of comparing molecular attributes. Very often a positive and a negative potential of a certain energy is used to describe structures. The presentation of potentials is based upon a charge calculation and may be used to find a suitable alignment of a training set of biologically active molecules. The latter can be realized on quite different quality levels. There are, for example, algorithms that perform well in calculations for simple carbohydrates, but are incapable of handling aromatic structures. Unfortunately these algorithms do not always signal their incapability if an aromatic system is to be calculated. A result is obtained, an isopotential surface is calculated, and a graph created. With that, an attempt is made to derive structure–activity relationships—the second trap comes next.

The training set that is selected, represents of course a drastic reduction of the parameter space. You may hope to receive a most possible representative distribution of the attributes by careful selection, but you are never sure. Thus, the correlations originate from the coincidential reciprocal completion of two errors, which relate back to the uncritical selection of methods and data sets.

1.3 What are Models Used for?

Models in science have different natures. They serve first of all to *simplify*; that means limiting of analysis to the phenomena that are believed to be the most important. Secondly, models serve as *didactical illustration* of very complicated circumstances, which are not easily accessible. Here, it must be taken into account in any explanation that the model does not show complete reality. A third model is that of *mechanical analogies*. These benefit from the fact that the laws of classical mechanics are completely defined, for example Hooke's law.

Model building of this kind plays a decisive role in the development of uniform theories. It is their special feature that it is not presupposed that the models reflect reality, but that first of all a structural similarity of two different fields is supposed. This is, for example, the presumption that the behavior of bonds in a molecule corresponds partly to springs, as described by Hooke's law. These mechanical analogy models have very successfully expanded theories, because the validity of a theory can in many cases be scrutinized experimentally, but the most important point is that predictions of new phenomena can be made.

These models are also often called *empirical.* Force fields belong to this class. The benefit of empiric models is that their parameters are optimized on reality. The "mechanization" does not provide explicit information from the non-mechanical contributions, but by empirical correction the non-mechanical contributions are convoluted in some way. That is why empirical models often are very close to reality.

Finally, the fourth type of model is *mathematical modeling.* These models serve for the simulation of processes, as for instance the kinetic simulation of a chemical reaction step in an enzyme. By suitable choice of parameters, kinetic simulations of real processes can be performed.

1.4 Molecular Modeling Uses All Four Types for Model Building

Didactical models are used for the combined representation of structure and molecular properties. In the case of *small molecules* the graphical representation of results from quantum chemical calculations or from the representation of the mobility of flexible ligands such as peptides. In the case of *proteins,* the structure itself is already a complex problem. Interactions of ligand and protein can also be studied with didactical models. It is already clear, that the different types of models are overlapping. Mechanical analogies, as well as reductions, aim at simplifying essential parts of the objects under study and are typical applications of molecular modeling.

1.5 The Final Step is *Design*

Design is perhaps the most essential element of all. Molecular modeling creates its own world, which is connected with reality by one of the four model types. Within this world—which exists in the computer—extrapolations can be made because, in contrast to the "real" world, a completely deterministical universe is created. Based on the analytical description of the system, the possibility is available to design inhibitors in advance of the synthesis and for them to be tested in a virtual computer experiment.

With that final design step, the circular course of a scientific study is completed. The study does not simply remain an analytical description of a system, which has been devised in "clockwork" fashion, but goes further by reassembling the system's parts. Molecular design creates a realization for our understanding—that a system could be more than simply the sum of its parts. This is especially effective for biological systems within which drug design is confronted by preference.

The design step itself actually is not as straightforward, even in the virtual world, as would be desirable. As *Gulliver* learns on his visit to the academy of Lagado, there is a machine, which at some time will have written every important scientific book of the world by a systematic combination of letters and words. *Jonathan Swift's* wonderful science fiction of the 18th century gives us at once the main problem: the time span of human beings is not large enough to test all possibilities. There has to be an intelligent algorithm to obtain the correct solutions. In the case of *Gullivers Travels,* Swift is somebody who introduces an additional criterion of quality. This is based on knowledge, experience, and is able to reject combinations of words and sentences: the human–machine network. Actually, Swift introduces such a criterion in the person of the professor who gives orders to his students, who serve the machines and decides after every experiment upon the result, e.g. lets the combination of words enter the book. Unfortunately, the experimenter himself is not defined qualitatively in Swift's novel; that is Swift's irony in *Gullivers Travels*. Hence, the result depends not only on an error-free function of the machine, but on the quality of its user! (Fig. 1)

The same problem is presented to us in the artificial world of modeling. Systematic exploration of properties is only possible for small numbers. Because of the combinatorics the system "explodes" after only a few steps. Flexibility studies on peptides give us a correct example. The change from four torsion angles to five or six increases the number of possible conformations from some thousands to several billions.

For the design of a ligand the situation becomes more complex. It demands a most intelligent restraint by suitable experiments, intuition or knowledge. Here also the quality of the human–machine network plays a decisive role. Fully automatic design systems seem to be like a Swift prediction machine in Gulliver's visit to the academy of Lagoda.

1.6 The Scope of the Book

The scope of this book is to provide support for the beginner. The recognition of principal concepts and their limitations is important to us—more important even than a complete presentation of all available algorithms, programs and data banks. As with all areas associated with computer techniques the technical development in this area has been more than exponential. Almost every day, new algorithms are offered on the network, suitable for comparison of protein sequences or for searching of new data banks, etc. The user has no other possibility to judge their quality than to use the programs and to explore their limitations.

He or she must know, therefore, that energy-minimizing in vacuo does not make sense in any case for the analysis of the interaction geometry of a ligand. He or she also has to know

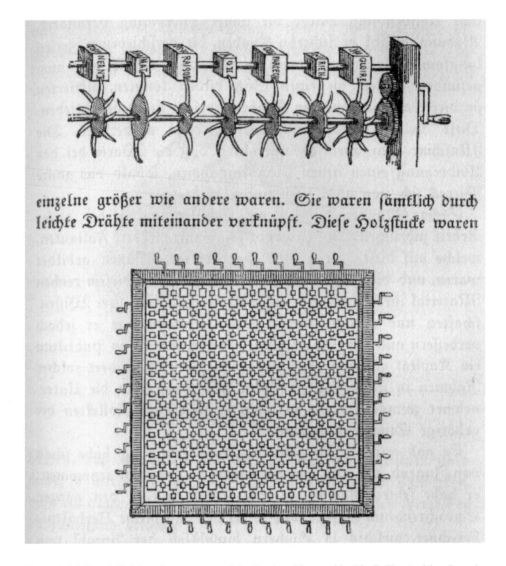

einzelne größer wie andere waren. Sie waren sämtlich durch leichte Drähte miteinander verknüpft. Diese Holzstücke waren

Figure 1. J. J. Grandville's imaginary concept of the "book-writing machine" in Gulliver's visit to Lagoda.

that a protein can not be simply folded up from a linear polypeptide chain. It must be realized that there is alternative or multiple binding mode: inhibitors binding to an enzyme show alternative binding geometries in the active site, even within a set of analogs. Very small changes in the molecular structure could provoke another orientation of the ligand in the active site. It is not necessarily true that a structure-orientated superposition would be better than an intuitive one or even one, which is oriented by steric or electrostatic surface properties.

Today's modeling in essence goes far beyond the example of Lukrez. Modeling is no longer on the level of analytic description of properties or correlations. It is much more than the creation of "colored pictures"—it also introduces us to systematic thinking. It even *demands* systematic thinking in order to avoid too many simple applications and to keep in mind the limitations of the methods.

Here we also want to provide support. By describing our own experiences with molecular design in two examples, one for "small molecules" (ligands) and another for "big molecules" (proteins), we aim to encourage the beginner to a critical engagement, hopefully without demotivation.

2 Small Molecules

2.1 Generation of 3D Coordinates

When starting a molecular modeling study the first thing to do is to generate a model of the molecule in the computer by defining the relative positions of the atoms in space by a set of cartesian coordinates. A reasonable and reliable starting geometry essentially determines the quality of the following investigations. It can be obtained from several sources. The three basic methods for generating 3D molecular structures are

1. use of X-ray crystallographic databases,
2. compilation from fragment libraries with standard geometries, and
3. simple drawing of 2D-structures using an approach called 'sketch'.

2.1.1 Crystal Data

First we will focus on the use of X-ray data for molecular building. The most important database for crystallographic information studying small molecules is the Cambridge Crystallographic Database [1]. This database contains experimentally derived atomic coordinates for organic and inorganic compounds up to a size of about 500 atoms and is continuously updated. The Cambridge Crystallographic Data Centre leases the database as well as software for searching the database and for analyzing the results. The output of the database search is a simple, readable file containing the 3D-structural information about the molecule of interest. This data file can be read by most of the commercial molecular modeling packages [e.g. 2, 3].

The atomic coordinates listed in the database are converted automatically to cartesian coordinates when reading the file into the modeling program. Subsequently the structure can be displayed by molecular graphics and studied in its 3D shape.

In general, small molecule X-ray structures are very well resolved but there is no guarantee for the accuracy of the data. The localization of hydrogen atoms always is a problem because they are difficult to observe by X-ray crystallography. The principle of the X-ray method is the scattering of the X-rays by the electron cloud around an atom. Because hydrogen atoms have only one electron, their influence on X-ray scattering is low and they are normally disregarded in structure determination. But of course hydrogen positions can be appointed on the basis of collected knowledge on standard bond lengths and bond angles. According to this procedure all bond lengths involving hydrogen atoms are usually not very specific. Before using the information from the X-ray database it is therefore advisable to check the atomic coordinates, bond lengths and bond angles for internal consistency. The following points especially should be clarified before starting any work with a X-ray structure:

1. are the atom types correct
2. are the bond lengths and bond angles reasonable
3. are the bond orders correct

and in case of chiral molecules,

4. do the data correspond to the correct enantiomer?

After taking care of these details the molecule can be saved in a molecular data file. The organization, extension name, format, and the information contained in the file are program-dependent.

It should be kept in mind that the crystalline state geometry of a molecule is subject to the influence of crystal packing forces. Therefore bond lengths and bond angles can differ from theoretical standard values. Furthermore, the solid state structure corresponds to only one of perhaps many low-energy conformations accessible to a flexible molecule and is always affected by the neighbor molecules in the crystal unity cell and sometimes also influenced by solvent molecules in the crystal. Other energetically allowed conformations must be explored by a conformational analysis eventually to reveal conformations of biological relevance. Also, knowledge of the most stable conformation called the *global energy minimum structure* is important to allow the evaluation of probabilities for conformers with higher energy content. Procedures for this purpose are described in section 2.2.

2.1.2 Fragment Libraries

The second very common building method is the construction of molecules from pre-existing fragment libraries. This is the method of choice when there is no access to crystallographic databases or if X-ray data for the desired structures are not available. Almost all commercial molecular modeling programs nowadays offer the possibility to construct molecules using fragment libraries.

Fragment libraries can be utilized like an electronic 3D structure tool kit, which is easy to handle. Because of the preoptimized standard geometries of all entries in the fragment pool resulting 3D structures already have an acceptable geometry. In most cases only torsion angles have to be cleared to avoid atom overlapping or close van der Waals contacts. Problems may arise with fused ring systems because of the different ways in which saturated rings can be joined to each other. To solve this problem it is recommended wherever possible to refer to X-ray data or to experimental data of comparable ring systems in order to select the correct ring connection.

Each atom in any arbitrary structure carries characteristic features which are defined by the so-called atom type. Properties distinguishing between different atoms in molecular modeling terms are for example hybridization, volume, etc. The corresponding parameters define the particular atom type. All atomic parameters taken collectively represent the atomistic part of a force field. On pre-existing fragments selected from libraries the atom types of course are already defined and in general are correct. In many cases, however, the decision as to which atom type will be appropriate is less easy to take. We will discuss this problem on the example of *N*-acetylpiperidine.

If *N*-acetylpiperidine is generated from the fragment library using a piperidine ring and an acetyl residue the nitrogen atom in the piperidine is defined as sp³ nitrogen atom type with tetrahedral geometry. But if this nitrogen is connected with the acetyl residue it also can be considered as an amide nitrogen atom demanding planar trigonal sp² geometry. In such a case the correct decision can only be made by either comparing the geometry obtained from the building routine with X-ray data or performing a quantum mechanical calculation for the structural element of interest in order to get a reliable geometry. Fig. 1 shows the results of a semiempirical and an ab initio calculation in comparison with force field geometries and the crystal structure of *N*-acetyl-piperidine-2-carboxylic-acid [4].

While the sp³ nitrogen atom of the force field structure bears a tetrahedral geometry the crystal structure and the quantum chemically calculated geometries indicate an almost planar nitrogen atom. To avoid errors in subsequent calculations the nitrogen atom has to be assigned an atom type with planar geometry.

Another problem occuring when building substituted saturated ring systems is the correct conformation of the cycle, because it may be influenced by the substituents. Cyclohexane is one of the most detailed studied cyclic molecules in organic chemistry. The different possible conformations and the energy barriers separating them have been the subject of many investigations [5, 6]. There is no doubt that the chair form is the most stable conformation of

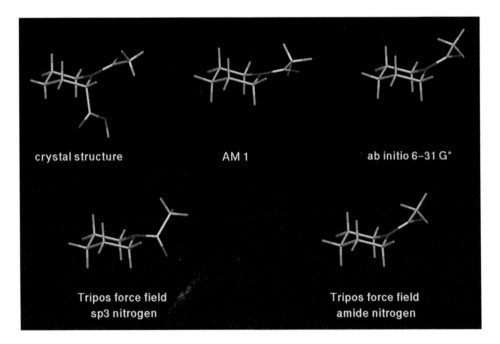

Figure 1. The geometry of the amide group in *N*-acetyl piperidine depends crucially on the method used as well as the atom types employed for optimization. For comparison the crystal structure of piperidine-2-carboxylic-acid is shown in the upper left. The color code: carbon = white, oxygen = red, nitrogen = blue, hydrogen = cyan, sulfur = yellow, halogens = green, is used throughout this book.

this molecule. For monosubstituted cyclohexane this still holds true. The preferred position of any substituent is found to be the equatorial one. The energy difference determined between the equatorial and axial position is not very distinct for small substituents but is increasing for larger groups [7]. Therefore it is necessary and advisable always to check the results of structure building from fragment libraries in comparable situations with experimental data.

2.1.3 Sketch Approach

The third method of structure generation is the so-called sketch approach. When using this routine the mouse pointer functions as a simple pencil to draw a 2D formula of the molecule on the computer screen. Sometimes a very limited number of small molecular standard fragments is already available from a library and can be used as starting points. When finishing the drawing process the 2D picture on the screen is converted into 3D information. Because of this procedure the setting of correct atom types should be watched especially carefully. Since the sketch approach is a very simple method the resulting geometries in general are not very satisfying. Therefore a rough geometry optimization is performed automatically at the end of each sketch operation in order to relax the molecular geometry.

References

[1]   Cambridge Structural Database, Dr. Olga Kennard, F.R.S., Cambridge Crystallographic Data Centre, 12 Union Road, Cambridge CB2 1EZ, U.K.
[2]   SYBYL, Tripos Associates, St. Louis, Missouri, USA.
[3]   INSIGHT/DISCOVER, Biosym Technologies Inc., San Diego, California, USA.
[4]   Rae, I. D., Raston, C. L., and White, A. H. *Aust. J. Chem.* **33**, 215 (1980).
[5]   Bucourt, R. The torsion angle concept in conformational analysis. In: *Topics in Stereochemistry*, Vol. 8. Eliel, E. L., and Allinger, N. L. (Eds.). Wiley: New York; 159–224 (1974).
[6]   Shopee, C. W. *J. Chem. Soc.* **II**, 1138–1151 (1946).
[7]   Hirsch, J. A. Tables of conformational energies. In: *Topics in Stereochemistry*, Vol. 1. Eliel, E. L., and Allinger, N. L. (Eds.). Wiley: New York; 199–222 (1967).

2.2 Computational Tools for Geometry Optimization

2.2.1 Force Fields

Molecular structures generated using the procedures described in the previous section should always be geometry optimized to find the individual energy minimum state. This is normally done by applying a molecular mechanics method. The expression "molecular mechanics" is used to define a widely accepted computational method employed to calculate molecular geometries and energies.

Unlike quantum mechanical approaches the electrons and nuclei of the atoms are not explicitly included in the calculations. Molecular mechanics considers the atomic composition of a molecule to be a collection of masses interacting with each other via harmonic forces. As a result of this simplification molecular mechanics is a relatively fast computational method practicable for small molecules as well as for larger molecules and even oligomolecular systems.

In the framework of the molecular mechanics method the atoms in molecules are treated as rubber balls of different sizes (atom types) joined together by springs of varying length (bonds). For calculating the potential energy of the atomic ensemble use is made of Hooke's law. In the course of a calculation the total energy is minimized with respect to atomic coordinates where:

$$E_{tot} = E_{str} + E_{bend} + E_{tors} + E_{vdw} + E_{elec} + ... \tag{1}$$

where E_{tot} is the total energy of the molecule, E_{str} is the bond-stretching energy term, E_{bend} is the angle-bending energy term, E_{tors} is the torsional energy term, E_{vdw} is the van der Waals energy term, and E_{elec} is the electrostatic energy term.

Molecular mechanics enables the calculation of the total steric energy of a molecule in terms of deviations from reference "unstrained" bond lengths, angles and torsions plus non-bonded interactions. A collection of these unstrained values, together with what may be termed force-constants (but in reality are empirically derived fit parameters), is known as the *force field*. The first term in Eq. (1) describes the energy change as a bond stretches and contracts from its ideal unstrained length. It is assumed that the interatomic forces are harmonic so the bond-stretching energy term can be described by a simple quadratic function given in Eq. (2):

$$E_{str} = \tfrac{1}{2} k_b (b-b_0)^2 \tag{2}$$

where k_b is the bond-stretching force constant, b_0 is the unstrained bond length, and b is the actual bond length.

In more refined force fields a cubic term [1], a quartic function [2–4], or a Morse function [5] has been included.

Also for angle bending mostly a simple harmonic, spring-like representation is employed. The expression describing the angle-bending term is shown in Eq. (3):

$$E_{bend} = \frac{1}{2} k_{\theta} (\theta - \theta_0)^2 \tag{3}$$

where k_{θ} is the angle-bending force constant, θ_0 is the equilibrum value for θ, and θ is the actual value for θ.

A common expression for the dihedral potential energy term is a cosine series, as Eq. (4):

$$E_{tors} = \frac{1}{2} k_{\varphi} (1 + \cos(n\varphi - \varphi_0)) \tag{4}$$

where k_{φ} is the torsional barrier, φ is the actual torsional angle, n is the periodicity (number of energy minima within one full cycle), and φ_0 is the reference torsional angle (the value usually is 0° for a cosine function with an energy maximum at 0° or 180° for a sine function with an energy minimum at 0°).

The van der Waals interactions between not directly connected atoms are usually represented by a Lennard-Jones potential [6] (Eq. 5).

$$E_{vdw} = \Sigma \frac{A_{ij}}{r_{ij}^{12}} - \frac{B_{ij}}{r_{ij}^6} \tag{5}$$

where A_{ij} is the repulsive term coefficient, B_{ij} is the attractive term coefficient, and r_{ij} is the distance between the atoms i and j.

This is one form of the Lennard-Jones potential but there exist several modifications of this term used in the different force fields. An additional function is used to describe the electrostatic forces. In general it is made use of the Coulomb interaction term (Eq. 6).

$$E_{elec} = \frac{1}{\varepsilon} \frac{Q_1 Q_2}{r} \tag{6}$$

where ε is the dielectric constant, Q_1, Q_2 are atomic charges of interacting atoms, and r is the interatomic distance.

Charges may be calculated using the methods described in section 2.4.1.1 or are implemented in some of the force fields [2–4] as empirically derived parameter sets.

Some force fields also include cross terms, out of plane terms, hydrogen bonding terms etc. and use more differentiated potential energy functions to describe the system. As force fields are varying in their functional form not all can be discussed here in detail but they have been subject of excellent reviews [7, 8].

The basic idea of molecular mechanics is that the bonds have "natural" lengths and angles. The equilibrium values of these bond lengths and bond angles and the corresponding force constants used in the potential energy function are defined in the force field and will be denoted as *force field parameters*. Each deviation from these standard values will result in increasing total energy of the molecule. So, the total energy is a measure of intramolecular strain relative to a hypothetical molecule with ideal geometry. By itself the total energy has no physical meaning.

The objective of a good and generally employable force field is to describe as many as possible different classes of molecules with reasonable accuracy. The reliability of the molecular mechanics calculation is dependent on the potential energy functions and the quality of the parameters incorporated in these functions. So, it is easy to understand that a calculation of high quality can not be performed if parameters for important geometrical elements are missing. To avoid this situation it is necessary to choose a suitable force field for a particular investigation.

Several force fields have been developed to examine a wide range of organic compounds and small molecules [1–4, 9] while other programs contain force fields primarily for proteins and other biomolecules [10–12]. If parameters for particular atom types are missing it is unavoidable to add the missing data to the force field [13–15].

2.2.2 Geometry Optimization

As already mentioned almost certainly the generated 3D model of a given molecule does not have ideal geometry; therefore, a geometry optimization must be performed subsequently. In the course of the minimization procedure the molecular structure will be relaxed. As can be deduced from the example presented in Fig. 1 and Table 1 the internal strain in structures obtained from crystal data is mainly influenced by small deviations from the "ideal" bond lengths. Therefore above all the corresponding energy terms (bond-stretching term, angle-bending term) are altered in course of a force field optimization. Despite the remarkable change in energy content torsional angles are effected only to a lesser extent. This is a clear indication to the well-known observation that in crystals almost exclusively low-energy conformations are found. It also should be realized that crystal structures are by no means

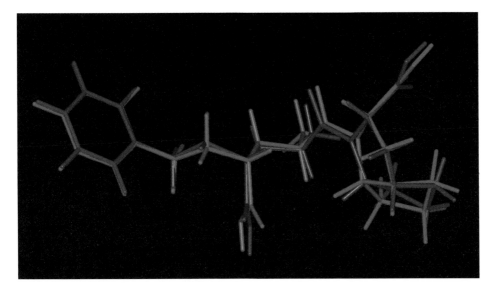

Figure 1. Superposition of the crystal structure (red) and force-field-optimized geometry (green) of the angiotensin-converting enzyme inhibitor ramiprilate.

Table 1. Force field energy terms for the ramiprilate molecule before and after geometry optimization

Structure	Energy (kcal mol^{-1})
Crystal	
Bond-stretching energy	179.514
Angle-bending energy	15.693
Torsional energy	17.230
Out-of-Plane-bending energy	0.043
1-4 van der Waals energy	18.538
van der Waals energy	−3.839
Total energy	227.178
Optimized	
Bond-stretching energy	0.982
Angle-bending energy	10.372
Torsional energy	14.335
Out-of-Plane-bending energy	0.011
1-4 van der Waals energy	4.791
van der Waals energy	−7.822
Total energy	22.669

"bad" geometries. As can be easily deduced from Fig. 1 the distortion of the crystal structure when compared with the relaxed geometry of the force field structure in terms of geometry differences is only very subtle. This fact can be interpreted also in the sense that large variations in geometry are not to be expected when different well-parameterized force fields are applied. In the case considered here the individual but real crystal packing of ramiprilate is compared to the well-known general Tripos force field.

Before starting a geometry optimization, bad van der Waals contacts should be removed because the minimum energy geometry at the end of the optimization will depend on the starting geometry [7].

Several advantages like speed, sufficient accuracy and the broad applicability on small molecules as well as on large systems have established the force field geometry optimization as the most important standard method. Because of the complexity and the demanding computational costs quantum mechanical methods should be reserved for special problems which will be discussed later.

We will now focus on some common energy minimization procedures used by molecular mechanics. It is important to note that the minimization algorithms only find local minima on the potential energy surface but not implicitly the global energy minimum.

2.2.3 Energy-Minimizing Procedures

The energy minimization methods can be divided into two classes: the first-derivative techniques like steepest descent, conjugate gradient and Powell; and the second-derivative methods like the Newton–Raphson and related algorithms.

2.2.3.1 Steepest Descent Minimizer

The steepest descent minimizer uses the numerically calculated first derivative of the energy function to approach the energy minimum. The energy is calculated for the initial geometry and then again when one of the atoms has been moved in a small increment in one of the directions of the coordinate system. This process will be repeated for all atoms which finally are moved to new positions downhill on the energy surface [7]. The procedure will stop if the predetermined minimum condition is fulfilled. The optimization process is slow near the minimum, so the steepest descent method is often used for structures far from the minimum. It is the method most likely to generate low-energy structures of poorly refined crystallographic data or to relax graphically built molecules. In most cases the steepest descent minimization is used as a first rough and introductory run followed by a subsequent minimization employing a more advanced algorithm like conjugate gradients.

2.2.3.2 Conjugate Gradient Method

The conjugate gradient method accumulates the information about the function from one iteration to the next. With this proceeding the reverse of the progress made in an earlier iteration can be avoided. For each minimization step the gradient is calculated and used as additional information for computing the new direction vector of the minimization procedure. Thus, each successive step continually refines the direction towards the minimum. The computational effort and the storage requirements are greater than for steepest descent but conjugate gradients is the method of choice for larger systems. The greater total computational expense and the longer time per iteration is more than compensated by the more efficient convergence to the minimum achieved by conjugate gradients.

The Powell method is very similar to conjugate gradients. It is faster in finding convergence and is suitable for a variety of problems, but one should be careful when using the Powell algorithm because torsion angles may sometimes be modified to a dramatic extent. So, the Powell method is not practicable for energy minimization after a conformational analysis because the located low-energy conformations will be altered in an undesired manner. It is advisable to perform a conjugate gradient minimization in this situation.

2.2.3.3 Newton–Raphson Minimizer

The Newton–Raphson minimizer as a second-derivative method uses, in addition to the gradient, the curvature of the function to identify the search direction. The second derivative is also applied to predict where the function passes through a minimum. The efficiency of the Newton–Raphson method increases as convergence is approached. The computational effort and the storage requirements for calculating larger systems are disadvantages of this method. For structures with high strain the minimization process can become instable, so the application of this algorithm is mostly limited to problems where rapid convergence from a preoptimized geometry to an extremely precise minimum is required. For some more detailed information about the optimization methods see [16, 17].

It can be summarized that the choice of the minimization method depends on two factors—the size of the system and the current state of the optimization. For structures far from minimum, as a general rule, the steepest descent method is often the best minimizer to use for the first 10–100 iterations. The minimization can be completed to convergence with conjugate gradients or a Newton–Raphson minimizer. To handle systems that are too large for storing and calculating a second-derivative matrix the conjugate gradient minimizer is the only practicable method. The minimization procedure will continue until convergence has been achieved.

There are several ways in molecular minimization to define convergence criteria. In non-gradient minimizers like steepest descent only the increments in the energy and/or the coordinates can be taken to judge the quality of the actual geometry of the molecular system. In all gradient minimizers, however, atomic gradients are used for this purpose. The best procedure in this respect is to calculate the root mean square gradients of the forces on each atom of a molecule. It is advisable also always to check the maximum derivative in order to detect unfavorable regions in the geometry. There is no doubt about the quality of a minimum geometry if all derivatives are less than a given value. The specific value chosen for example for the maximum derivative depends on the objective of the minimization. If a simple relaxation of a strained molecule is desired, a rough convergence criterion like a maximum derivative of 0.1 kcal mol^{-1} Å^{-1} is sufficient while for other cases convergence to a maximum derivative less than 0.001 kcal mol^{-1} Å^{-1} is required to find a final minimum.

The choice of the convergence criteria should be a balance between attaining reasonable accuracy in determining the minimum structure and avoiding unnecessary computations when no further progress can be realized [17].

2.2.4 Use of Charges, Solvation Effects

Molecular mechanics calculations are usually carried out under vacuum conditions ($\varepsilon = 1$). For unpolar hydrocarbons the effect of the explicit inclusion of solvent as compared with gas phase calculations is negligible. The investigation of molecules containing charges and dipoles however requires the consideration of solvent effects [7]; otherwise conformations mainly influenced by strong electrostatic interactions would be overestimated. The force field will try to maximize the attractive electrostatic interaction, resulting in energetically strongly preferred but unrealistic low-energy conformations of the molecule. This can be prevented by employing the corresponding solvent dielectric constant [18]. For example, in water ε amounts to 80. In contrast to macromolecules, the electrostatic field of small molecules is considered to be homogeneous; therefore the use of an uniform dielectric constant in principle is allowed. Experimentally determined dielectric constants for a large number of solvents may be found in the literature and can be applied for a correct treatment of the Coulombic term of solvated molecules.

A very simple but effective way to treat the problem of charges and solvation in the course of a molecular mechanics optimization is to perform the calculation without taking charges into account. This very often yields acceptable results and is especially recommended if the results of a conformational analysis are to be minimized because usage of charges may markedly alter the conformation by electrostatic interactions. Consideration of charges always is necessary if hydrogen bonding phenomena are to be described.

The strength of the electrostatic interaction decreases with r^{-1}. Therefore, in some force fields the dielectric constant can be chosen to be distance-dependent in order to simulate the effect of displacement of solvent molecules in course of the approach of a ligand molecule to a macromolecular surface. This is of particular value if a conformational analysis is part of a pharmacophore search.

Whenever possible experimental data should be used for testing results from theoretical calculations. Above all, NMR data have become a valuable tool in this respect. Since most of the available NMR data have been obtained in chloroform or similar organic solvents, the explicit inclusion of the corresponding dielectric constant in the Coulombic term of a force field leads to an improved agreement with experimental results.

Consideration of the dielectric constant is one possibility to simulate solvent effects. An alternative way is to create a solvent box around the molecule containing discrete solvent molecules. The additional computational effort and the limitations in regard to the limited number of solvents that can be used in most of the available force fields are severe disadvantages of this method.

2.2.5 Quantum Mechanical Methods

Quantum mechanical methods also must be discussed, at least in brief, because they are very valuable additional tools in computational chemistry. In general, properties like molecular geometry and relative conformational energies can be calculated with high accuracy for a broad variety of structures by a well-parametrized general force field. However, if force field parameters for a certain structure are not available quantum chemical methods can be used for geometry optimization. In addition, the calculation of transition states or reaction paths as well as the determination of geometries influenced by polarization or unusual electron distribution in a molecule is the domain of quantum mechanical calculations. Their disadvantages relative to other methods are the computational costs and the limitation to rather small molecules. So, the use of quantum mechanical methods should be reserved for the treatment of special problems. The objective in this context is not to discuss the quantum mechanical methods from a theoretical perspective but to give some practical hints for the application of semiempirical or ab initio programs. The reader's interest may be drawn to many books and reviews on this subject to gain more insight into the theoretical aspects of these methods [19–22].

2.2.5.1 Ab initio Methods

Unlike molecular mechanics and semiempirical molecular orbital methods ab initio quantum chemistry is capable of reproducing experimental data without employing empirical parameters. Therefore, the application of ab initio calculations is especially favored in situations in which little or no experimental information are available.

The quality of an ab initio calculation depends on the basis set used for the calculation [23, 24] and the computational method employed. A wrong choice of the basis set can render the results of extremely time-consuming calculations meaningless. The decision which basis set should be used is related to the objective of the calculation and the molecules to be studied. It should be kept in mind that even a large basis set is not always a guarantee for agreement with experimental data [25].

Only the most commonly applied basis sets will be discussed here. The STO-3G (Slater type orbitals approximated by three Gaussian functions each) basis set has been frequently used in the past and is the smallest basis set that can be chosen. This minimal basis set consists of the smallest number of atomic orbitals necessary to accommodate all electrons of the atoms in their ground state, assuming spherical symmetry of the atoms.

In more recent ab initio calculations the split-valence basis sets have become quite popular. In these the valence orbital shells are represented by an inner and outer basis function. In this way more flexibility in describing the residence of the electrons has been attained [26]. The split-valence basis sets represent a progress over the STO-3G basis set, and the 3-21G, 4-31G, and 6-31G basis sets are widely used in ab initio calculations. They differ only in the number of primitive Gaussians used in expanding the inner shell and first contracted valence function [25]. 4-31G for example means that the core orbitals consist of four and the inner and outer valence orbitals of three and one Gaussian functions, respectively.

The next level of improvement is the introduction of polarization basis sets. To all non-hydrogen atoms d orbitals are added to allow p orbitals to shift away from the position of the nucleus leading to a deformation (polarization) of the resulting orbitals. This adjustment is particulary important for compounds containing small rings [26]. The polarization basis sets are marked by a star, e.g. 6-31G*. This basis set uses six primitive Gaussians for the core orbitals, a three/one split for the s and p valence orbitals, and a single set of six d functions (indicated by the asterisk).

For a more detailed description of the basis sets the reader is directed to books and reviews on this subject [22, 25].

Unfortunately there is no general rule for choosing an adequate basis set. The level of calculation depends on the desired accuracy and the molecular properties of interest. A geometry optimization of a simple molecule with moderate size reasonably can be performed using a 3-21G basis set. For other problems, however, this degree of sophistication may not be sufficient. If the geometry of the molecule is influenced by polarization effects, electron delocalization or hyperconjugative effects a 6-31G* or higher basis set is necessary to include the d orbitals as already mentioned (Fig. 2).

In spite of the rapid development in computer technology, high level ab initio calculations still can not always be performed. A common way to overcome the problem of excessive computational requirements is the use of a 3-21G basis set to optimize the geometry parameters and then to compute the wavefunction on the 6-31G* level. This procedure is often termed 6-31G*//3-21G calculation.

The use of higher basis sets does not automatically improve the accuracy of the calculated molecular properties of interest. In order to find a suitable level of calculation it is necessary to calibrate the method against experiment or testing the basis sets empirically to yield acceptable results.

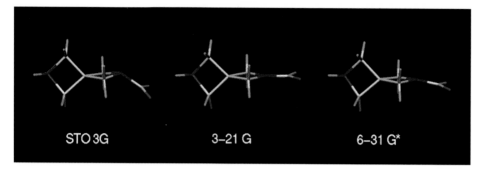

STO 3G   3–21 G   6–31 G*

Figure 2. This shows the final geometries of 2,6-diazaspiro[3.3]hept-2-yl-formamide after geometry optimization using different basis sets. The example clearly indicates the dependence of the resulting geometry on the applied basis set. The minimal basis set STO-3G and the 3-21G basis set yield very different geometries. The inclusion of d orbitals (6-31G*) leads to a structure reflecting the polarization effects and the ring tension more precisely. The resulting geometry of the amide nitrogen atom lies between tetrahedral and trigonal planar hybridization states.

2.2.5.2 Semiempirical Molecular Orbital Methods

The deep gap between molecular mechanics and the ab initio calculations is occupied by the semiempirical molecular orbital methods. They are basically quantum mechanical in nature but the main difference to ab initio methods is the introduction of empirical parameters in order to reduce the high costs of computer time necessary for explicit evaluation of all integrals. One-center repulsion integrals and resonance integrals are substituted by parameters fitted as closely as possible to experimenal data.

Another basic idea of the semiempirical approach is the consideration of the fact that most of the interesting molecular properties are mainly influenced only by the valence electrons of the corresponding atoms. Therefore only the valence electrons are taken into account, leading to a further reduction in computer time.

All the semiempirical methods apply the same theoretical assumptions, they only differ in the approximations beeing made [27]. Semiempirical methods like AM1 [28] and PM3 [29–31] provide a quite effective compromise between the accuracy of the results and the expense of computer time required. A calculation performed with AM1 or PM3 is able to reflect the experiment as effectively as an ab initio calculation using a small basis set. The advantage of semiempirical methods over ab initio calculations is not only that they are several orders of magnitude faster, but also that calculations for systems up to 200 atoms are possible with the semiempirical methods only. However, it is recommended to check one's results carefully. Like the choice of a wrong basis set in ab initio calculations, the lack of correct parameters in semiempirical studies can also lead to meaningless results. The quality of semiempirical methods for a wide range of molecules and the calculation of different properties has been subject of several reviews [28–31]. It should be noted that in general semiempirical methods may give erroneous results for the third-row elements.

References

[1] Allinger, N. L. *J. Am. Chem. Soc.* **99**, 8127–8134 (1977).

[2] Allinger, N. L., Yuh, Y. H., and Lii, J.-H. *J. Am. Chem. Soc.* **111**, 8551–8566 (1989).

[3] Lii, J.-H., and Allinger, N. L. *J. Am. Chem. Soc.* **111**, 8566–8576 (1989).

[4] Lii, J.-H., and Allinger, N. L. *J. Am. Chem. Soc.* **111**, 8576–8582 (1989).

[5] Morse, P. M. *Phys. Rev.* **34**, 57 (1929).

[6] Lennard-Jones, J. E. *Proc. Roy. Soc.* **106A**, 463 (1924).

[7] Burkert, U., and Allinger, N. L. *Molecular Mechanics*. ACS Monograph 177. American Chemical Society: Washington D. C. 1982.

[8] Dinur, U., and Hagler, A. T. New Approaches to Empirical Force Fields. In: *Reviews in Computational Chemistry*, Vol. 2. Lipkowitz, K. B., and Boyd, D. B. (Eds.). VCH: New York; 99–164 (1991).

[9] Clark, M., Cramer III, R.D., and Van Opdenbosch, N. *J. Comput. Chem.* **10**, 982–1012 (1989).

[10] Dauber-Osguthorpe, P., Roberts, V.A., Osguthorpe, D.J., Wolff, J., Genest, M., and Hagler, A.T. *Proteins: Structure, Function and Genetics* **4**, 31–47 (1988).

[11] Brooks, B. R., Bruccoleri, R. E., Olafson, B. D., States, D. J., Swaminathan, S., and Karplus, M. *J. Comput. Chem.* **4**, 187–217 (1983).

[12] van Gunsteren, W.F., and Berendsen, H.J.C. Molecular dynamics simulations: techniques and applications to proteins. In: *Molecular Dynamics and Protein Structure*. Hermans, J. (Ed.). Polycrystal Books Service: Western Springs, Illinois; 5–14 (1985).

[13] Hopfinger, A. J., and Pearlstein, R. A. *J. Comput. Chem.* **5**, 486–499 (1984).

[14] Maple, J.R., Dinur, U., and Hagler, A. T. *Proc. Natl Acad. Sci., U.S.A.* **85**, 5350–5354 (1988).

[15] Bowen, J. P., and Allinger, N. L. Molecular Mechanics: The Art and Science of Parameterization. In: *Reviews in Computational Chemistry*, Vol. 2. Lipkowitz, K. B., and Boyd, D. B. (Eds.). VCH: New York; 81–97 (1991).

[16] Press, W. H., Flannery, B. P., Teukolsky, S. A., and Vetterling, W. T. *Numerical Recipes in C*. Cambridge University Press: Cambridge 1988.

[17] Schlick, T. Optimization Methods in Computational Chemistry. In: *Reviews in Computational Chemistry*, Vol. 3. Lipkowitz, K. B., and Boyd, D. B. (Eds.). VCH: New York; 1–71 (1992).

[18] Eliel, E. L., Allinger, N. L., Angyal, S. J., and Morrison, G. A. *Conformational Analysis*. Wiley-Interscience: New York 1965.

[19] Pople, J. A. *Acc. Chem. Res.* **3**, 217 (1970).

[20] Hehre, W. J., Radom, L., Schleyer, P. v. R., and Pople, J. A. *Ab Initio Molecular Orbital Theory*. Wiley-Interscience: New York 1986.

[21] Szabo, A., and Ostlund, N. S. *Modern Quantum Chemistry: Introduction to Advanced Electronic Structure Theory*. McGraw-Hill: New York 1985.

[22] Clark, T. *A Handbook of Computational Chemistry: A Practical Guide to Chemical Structure and Energy Calculations*. Wiley-Interscience: New York 1985.

[23] De Frees, D. J., Levi, B. A., Pollack, S. K., Hehre, W. J., Binkley, S. J., and Pople, J. A. *J. Am. Chem. Soc.* **101**, 4085–4089 (1979).

[24] Davidson, E. R., and Feller, D. *Chem. Rev.* **86**, 681–696 (1986).

[25] Feller, D., and Davidson, E. R. Basis Sets for Ab Initio Molecular Orbital Calculations and Intermolecular Interactions. In: *Reviews in Computational Chemistry*, Vol. 1. Lipkowitz, K. B., and Boyd, D. B. (Eds.). VCH: New York; 1–43 (1990).

[26] Boyd, D. B. Aspects of Molecular Modeling. In: *Reviews in Computational Chemistry*, Vol. 1. Lipkowitz, K. B., and Boyd, D. B. (Eds.). VCH: New York; 321–354 (1990).

[27] Kunz, R. W. *Molecular Modelling für Anwender*. Teubner: Stuttgart 1991.

[28] Dewar, M. J. S., Zoebisch, E. G., Healy, E. F., and Stewart, J. J. P. *J. Am. Chem. Soc.* **107**, 3902–3909 (1985).

[29] Stewart, J. J. P. Semiempirical Molecular Orbital Methods. In: *Reviews in Computational Chemistry*, Vol. 1. Lipkowitz, K. B., and Boyd, D. B. (Eds.). VCH: New York; 45–81 (1990).

[30] Stewart, J. J. P. *J. Comput. Chem.* **10**, 209–220 (1989).

[31] Stewart, J. J. P. *J. Comput. Chem.* **10**, 221–264 (1989)

2.3 Conformational Analysis

Molecules are not rigid. The motional energy at room temperature is large enough to let all atoms in a molecule move permanently. That means that the absolute positions of atoms in a molecule, and of a molecule as a whole, are by no means fixed and that the relative location of substituents on a single bond may vary in the course of time. Therefore, each compound containing one or several single bonds is existing at each moment in many different so-called *rotamers* or *conformers*. The quantitative and qualitative composition of this mixture is permanently changing. Of course only the low-energy conformers are found to a large extent.

A transformation from one conformation to another is primarily related to changes in torsion angles about single bonds. Only minor changes of bond lengths and angles take place. The changes in molecular conformations can be regarded as movements on a multi-dimensional surface that describes the relationship between the potential energy and the geometry of a molecule. Each point on the potential energy surface represents the potential energy of a single conformation. Stable conformations of a molecule correspond to local minima on this energy surface. The relative population of a conformation depends on its statistical weight which is influenced not only by the potential energy but also by the entropy. As a consequence, the global minimum on the potential energy surface—the conformation which contains the lowest potential energy—does not necessarily correspond to the structure with the highest statistical weight (for a detailed description see [1]).

Well-known examples for multiple conformations of molecules are the staggered and eclipsed forms of ethane, the anti-*trans* and *gauche* forms of *n*-butane or the boat and chair forms of cyclohexane. The rotation about the C_{sp3}–C_{sp3} bond in the ethane molecule can be described by a sine-like curve potential function (Fig. 1). The energy minima, located at 60°, 180° and 300°, correspond to the staggered form, while the maxima, located at 120°, 240° and 360°, correspond to the eclipsed form of ethane. Because structures located at maxima on the potential energy function (or potential energy surface) are not viable normally, only the staggered form of ethane needs to be taken into account when physical or chemical properties are studied. This straightforward situation completely changes in the case of larger and more flexible molecules which exist at room temperature in several energetically accessible rotamers. For example, at room temperature approximately 70% of *n*-butane exist in the anti-*trans* form and 30% in the *gauche* form [2]. Thus, for a discussion of the physical behavior of this flexible aliphatic chain both the anti-*trans* and the *gauche* conformations have to be taken into account. The same is true for cyclic structures like cyclohexane, where the chair as well as the boat form must be regarded.

The biological activity of a drug molecule is supposed to depend on one single unique conformation hidden among all the low-energy conformations [3]. The search for this so-called bioactive conformation for sets of compounds is one of the major tasks in medicinal chemistry. Only the bioactive conformation can bind to the specific macromolecular environment at the active site of the receptor protein. Based on the information of the active conformation one may be able to design new agents for a particular receptor system. It is widely accepted that the bioactive conformation is not necessarily identical with the lowest-energy conformation. However, on the other hand it cannot be a conformation that is so high in energy that it is excluded from the population of conformations in solution (for a discussion of this aspect see

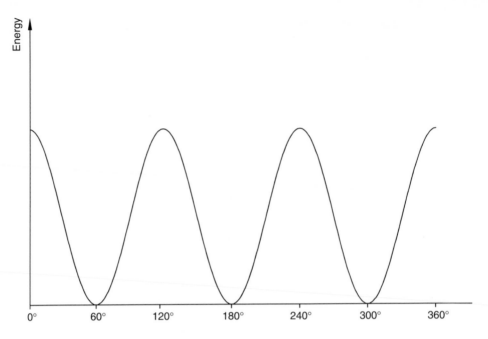

Figure 1. Sine-like potential energy curve of ethane shown as function of the dihedral angle.

[4]). Thus, the identification of low-energy conformations is an important part of understanding the relationship between the structure and the biological activity of a molecule.

Experimental techniques such as NMR only provide information on one or a few conformations of a molecule. A complete and exclusive overview of the conformational potential of molecules can be gained by theoretical techniques. Correspondingly a variety of theoretical methods for conformational analysis has been developed. Many applications are reported in the literature [5–12]. The most general methods for conformational analysis are those that are able to identify all minima on the potential energy surface. However, as the number of minima increases dramatically with the number of rotatable bonds, an exhaustive detection of all minima becomes a difficult and time-consuming task.

The time required for a conformational analysis depends also directly on the type of method used for the calculation of the energy. Conformational energies can be calculated either using quantum mechanical or molecular mechanical methods. Because the quantum mechanical calculations are very time consuming, they cannot be applied to large or flexible molecules. For that reason most of the conformational search programs use molecular mechanics methods for the calculation of energies as a standard. Apart from systematic search procedures we also will deal in this chapter with the use of Monte Carlo and molecular dynamics techniques for conformational analyses.

2.3.1 Conformational Analysis Using Systematic Search Procedures

The systematic search [6,7,13] is perhaps the most natural of all different conformational analysis methods. It is performed by varying systematically each of the torsion angles of a molecule in order to generate all possible conformations. If the angle increment is appropriately small the procedure yields a complete image of the conformational space of any molecule.

The step size which is normally used in a systematic search is 30°. That means, during a full rotation of 360°, 12 conformations are generated. In close neighborhood to the optimal value a smaller step size down to 5° may be necessary in order to determine the minimum position of a conformation exactly. The number of generated conformations depends on the step size, but also on the number of rotatable bonds. If n is the number of rotatable bonds, then the number of conformations increases with the nth power:

Number of conformations = $(360/\text{step size})^n$

If for example a systematic conformational search is performed for a molecule with six rotatable bonds and a step size of 30° is employed, the number of generated conformers amounts to 12^6 or 2 985 984 structures. This huge amount of data cannot be handled; it therefore has to be reduced.

The first step in data reduction is a van der Waals screening or "bump check". It is performed before the potential energy of the conformations will be exactly calculated. The screening procedure excludes all conformations where a van der Waals volume overlap of atoms not directly bound to each other is detected. The mathematical criterion for determining the validity of a conformation in this respect simply is the sum of the van der Waals radii of non-bonded atoms. The hardness of van der Waals spheres can be varied by specification of a so-called van der Waals factor. This multiplication constant controls the interpenetrability of atoms. A reduction of the van der Waals factor results in softening of contacts between non-bonded atoms, thereby increasing the number of valid conformations.

For the conformers remaining after the bump check the potential energy is calculated using a molecular mechanics method. In general the conformational energy is calculated neglecting electrostatic interactions, i.e. charges are not taken into account and the conformational analysis is performed in vacuo. The reasons for this procedure have been discussed in section 2.2.4. If the inclusion of electrostatic interactions into a conformational analysis is justified for a special case then the whole process becomes much more complex. Good-quality atomic charges are sensitive to the discrete spatial environment of the atoms and not only depend on the connectivity. Therefore, atomic charges which have been calculated for the initial conformation must be constantly updated after each modification of a torsion angle. In addition it would be necessary to mimick the effect of a solvent, which tones down the strong electrostatic interactions built up between charges in vacuo. Obviously this procedure would require a large amount of additional computer time, even for a small molecule. And what is even more noteworthy, the increase in complexity of the system does not produce a deeper insight into the conformational behavior of a molecule in solution, besides the fact that intramolecular interactions are diminished. The same result is obtained when charges are not considered and the analysis is performed in vacuo. Besides, in the active site of a receptor or enzyme the intramolecular contacts in ligands are also of minor importance.

When the conformational energies have been calculated for all conformers which survived the bump check another possibility to reduce the number of conformations is the use of an energy window. The underlying idea for applying an energy window is based on the fact that conformations containing much more energy than those close to the minimum are found in the conformer population only to a neglectable quantity, i.e. in our context it may be assumed that they do not have any importance for the biological activity of a particular molecule. The value for this energy window depends on the size of the studied molecule as well as on the applied force field. It may vary between 5 and 15 kcal mol^{-1} [11–15].

The resulting conformations, which have passed all filter methods should represent a complete ensemble of energetically accessible conformations for a particular molecule. However, in many cases the number still may be too large to allow a reasonable treatment. Many of the remaining conformations are very strongly related, because they only differ for example in a single rotor step. Obviously these can be combined to a common family with pronounced similarity. The description of the conformational properties of a molecule does not lack comprehensiveness if we only take the minimum conformer of each conformational family into further consideration. Several methods have been developed to execute the classification into conformational families [15–17]. The parameters used for this purpose are the torsion angles. The known classification methods differ in the procedure to associate the conformers to individual families. Another possibility to evaluate the large amount of data accumulated in course of a systematic conformational search is the application of statistical techniques like cluster or factor analysis. For a detailed discussion of these methods see [18].

The course of a systematic conformational analysis shall be demonstrated on a study performed in our group with two H_2-antihistaminic agents, tiotidine and ICI127032 (Fig. 2) [19]. It was performed using the SEARCH module within the molecular modeling package SYBYL [16].

As rotational increment a step size of 15° was chosen. Due to symmetry the methyl group of the cyanoguanidine system was only rotated in steps of 30° between 0° and 120°. The theoretical number of conformations, 3.98×10^7, was reduced using the van der Waals screening to 4.6×10^6, i.e. roughly 10% of the initial number still is valid after the bump check. The application of an energy window of 15 kcal mol^{-1} leads to a further reduction of 90%. Some 453 393 conformations were stored. Even this number cannot be handled in a reasonable way. Therefore, in a next step the conformations left were classified into families using the program IXGROS [17] which has been developed in our group. (The complete source code is listed in Appendix 1.) This finally yields 227 unique families which are represented by their respective minimum energy conformations. Although the reduction from 4.0×10^7 down to 227 conformations is very impressive, one has to submit that even the rather small number left is too large. There is no chance to decide which of the 227 conformers is the bioactive one, but this and only this is the question of interest. At this point a solution cannot be found if there do not exist rigid or at least semi-rigid congeners which in addition must be biologically active. It also must be proven that they bind to the same receptor site in an analogous mechanism. That is, as a rule, for finding the bioactive conformation of a flexible molecule, potent and more rigid compounds of the same series are needed. In the case of the H_2 antagonists the rigid and

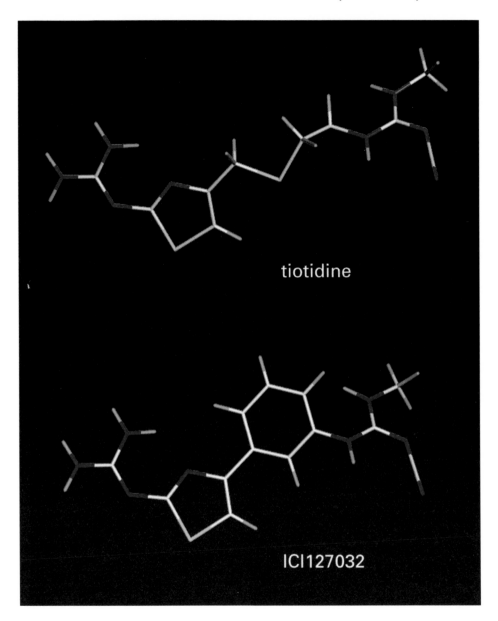

Figure 2. Molecular formulas of the histamine H_2 receptor antagonists tiotidine and ICI127032.

potent representative is ICI127032. After consideration of the small number of low-energy conformations of the rigid matrix and repeated use of IXGROS, eight unique families survived the procedure. These remaining conformations could be used successfully to determine the biologically active conformation of tiotidine (Fig. 3).

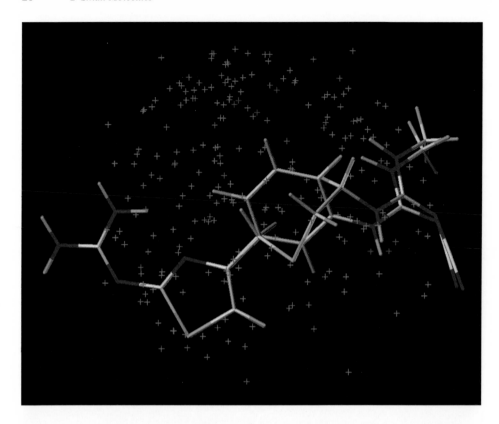

Figure 3. Representation of the results of the conformational analysis of tiotidine and ICI127032 (both displayed in a possible minimum energy conformation). The local minimum conformation representing the different conformational families are displayed by stars symbolizing the center of the cyanoguanidine end group of tiotidine and ICI127032. The resulting conformations of tiotidine are indicated by green stars, while the red-coloured stars mark the conformations derived for ICI127032. (The calculations have been performed using the SEARCH module within SYBYL 6.1 [16] and IXGROS [17]).

As we have discussed it is of advantage to include rigid molecules in a conformational search for a set of flexible congeners. The rigid and biologically potent derivatives are used as a matrix for all other members of the series. Marshall and colleagues [7] have extended this procedure by also including inactive rigid representatives. By doing this the conformational space can be further restricted and by the same token the time necessary for the search is reduced by orders of magnitude. This technique has become known as "Active analogue approach".

2.3.2 Conformational Analysis Using Monte Carlo Methods

A completely different path for searching conformational space is realized in the Monte Carlo or random search. Random search techniques are of a statistical nature [20]. At each stage of a Monte Carlo search the actual conformation is modified randomly in order to obtain a new one.

A random search starts with an optimized structure. At each iteration in the procedure, new torsion angles [11] or new cartesian coordinates [8,9] are assigned randomly. The resulting conformation is minimized using molecular mechanics and the randomization process is repeated. The minimized conformation is then compared with the previously generated structures and is only stored if it is unique. The random methods potentially cover all regions of conformational space, but this only is true if the process is allowed to run for a sufficiently long time. This may last extremely long because the probability to detect a new and unique conformation decreases dramatically depending on the growing number of conformers already discovered. However, even if the computation has been running very long, one cannot be certain that the conformational space has been completely covered. It is very important therefore to establish a means for testing the completeness of the analysis. This can be done efficiently by performing several runs in a parallel mode, each one starting with a different initial conformation. If the results are identical or nearly identical, then completeness can be assumed. Another measure of completeness is based on the recovery rate for each low-energy conformation, because the probabilistic process must reproduce it many times.

The main advantage of random search methods is that, in principle, molecules of any size can be successfully treated. In practice, however, highly flexible molecules often do not give converging results, because the volume of the respective conformational space is too large. Other useful applications for Monte Carlo search methods include investigations on cyclic systems, because ring systems in general are difficult to treat in systematic searches. The effectiveness of random search procedures shall be demonstrated on a practical example. Cycloheptadecane was studied using a variety of different methods including a random search method [12]. The combined results of the various procedures yielded a total amount of 262 different minimum conformations. None of the employed techniques succeeded in finding all 262 conformers, but one of the random search analyses nevertheless was able to detect 260 of them. It is therefore safe to comment that random search techniques are very suitable for conformational analyses of many types of molecules, but may require a large amount of computer time to ensure complete coverage of conformational space.

2.3.3 Conformational Analysis Using Molecular Dynamics

The systematic conformational search procedure is a valuable tool to determine the large number of minima on the potential energy surface associated with a flexible molecule. In principle, the generation of all allowed conformations can be realized and there is a high probability for the completeness of the conformational search. However, there are clear limitations in the applicability of this method. The multi-minima problem can only be solved for rather small molecules with a limited number of rotatable bonds.

As already mentioned in section 2.3.1 the systematic conformational search of a molecule with six rotatable bonds leads to serious problems in data handling due to the large number of generated conformers. Therefore the investigation of flexible molecules—like for example arachidonic acid (Fig. 4), which contains 15 rotors—is practically impossible. Even after applying several methods of data reduction the systematic conformational search for this molecule yielded 500 000 different conformations. The procedure was stopped automatically by the program due to data overflow, although the conformational space was not completely sampled at this point.

However, conformational analysis of the same molecule by a random search procedure will also be unreasonable because of the required computer time. For example, the investigation of cycloheptadecane—which is a more restricted molecule—used about 94 days of computer time on a Micro-Vax II computer [12].

Another rather difficult subject in this context is presented when saturated or partially saturated ring systems are to be treated in a systematic conformational analysis. In the course of the systematic process, bonds have to be broken in order to produce new attainable ring conformations. Efficiency and reliability of this procedure have been subject of several reviews [12, 14].

A very common strategy to overcome these problems is the use of molecular dynamics simulations for exploring conformational space. The aim of this approach is to reproduce the time-dependent motional behavior of a molecule. Molecular dynamics are based on molecular mechanics. It is assumed that the atoms in the molecule interact with each other according to the rules of the employed force field (as already described in section 2.2.1). At regular time intervals the classical equation of motion represented by Newton's second law is solved:

$$F_i(t) = m_i\, a_i(t) \tag{1}$$

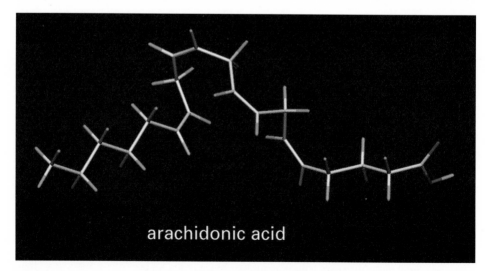

Figure 4. One energetically permitted conformation of arachidonic acid.

where F_i is the force on atom i at time t, m_i is the mass of atom i, and a_i is the acceleration of atom i at time t. The gradient of the potential energy function is used to calculate the forces on the atoms while the initial velocities on the atoms are generated randomly at the beginning of the dynamics run. Based on the initial atom coordinates of the system, new positions and velocities on the atoms can be calculated at time t and the atoms will be moved to these new positions. As a result of this a new conformation is created. The cycle will then be repeated for a predefined number of time steps. The collection of energetically accessible conformations produced by this procedure is called an *ensemble*.

The application of Newton's equations of motion is uniform in all different available molecular dynamics approaches, but they differ in the employed integration algorithms. Very common methods for integrating the equations of motion are the Verlet integrator [21] and algorithms like Beeman [22] and the leap-frog scheme [23] which are simple modifications of the Verlet algorithm. In the framework of this book a more extended discussion of the molecular dynamics theory is not intended but the interested reader is urged to study more detailed reviews on this subject [24–27].

Before employing molecular dynamics simulations for conformational analysis the reader's attention should be drawn to some special features of this method. Unlike the conservative geometry optimization procedures, molecular dynamics is able to overcome energy barriers between different conformations. Therefore it should be possible to find local minima other than the nearest in the potential energy surface. However, if the energy barrier is high or the number of degrees of freedom in the molecule is very large, then some of the existing conformers of the investigated system possibly are not reached. In view of the huge conformational space the completeness of the conformational search during the chosen simulation time is difficult to ensure.

To enhance conformational sampling a widely used tactics in molecular dynamics is to apply an elevated temperature to the simulation [27]. At high temperature the molecule is able to overcome even large energy barriers that may exist between some conformations and therefore the chance for completeness of a conformational search increases. It is self evident that the choice of a particular simulation temperature and simulation time depends closely on the molecule of interest.

One recent and comprehensive investigation can be used to demonstrate the dependence of conformational flexibility on the simulation temperature. The data and additional material were made available by courtesy of F. S. Jorgensen, Copenhagen (Denmark). A molecular dynamics simulation has been performed on the experimentally well-studied cyclohexane molecule using different start conformations and different simulation temperatures* (Fig. 5).

At 400 K the twist form of cyclohexane ($T_1 = 0$) which has been used as initial conformation, oscillates between different twist forms while at 600 K the molecule contains sufficient kinetic energy to convert to one of the chair conformations ($T_1 = 300$). Further increase of the temperature up to 1000 K yielded both chair as well as twist conformations

---

* Sybyl (version 6.0.3) from Tripos Associates Inc., St. Louis, U.S.A. Energy minimizations: Tripos force field, PM3 partial charges, dielectric constant $\varepsilon = 20$ and a convergence criterion of 0.005 kcal mol^{-1}Å^{-1}. MD simulations: 1000 ps at various temperatures with conservation of total energy, one conformation sampled per picosecond.

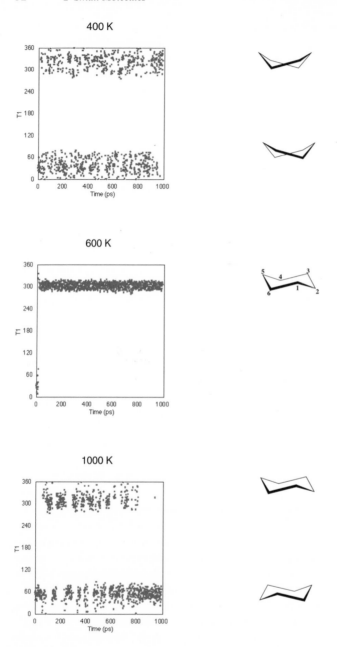

Figure 5. Variation of torsion angle T_1 (torsion angle $T_1 = C_1-C_2-C_3-C_4$) of cyclohexane for different simulation temperatures. At 400 K the molecule oscillates between different flexible twisted boat forms reflected by an extensive fluctuation of the observed torsion angle. Increasing the temperature to 600 K leads to one of the possible stable minima corresponding to one chair conformation. The dynamic simulation at 1000 K yields both chair conformations as well as the already observed twist and boat conformations.

($T_1 = 300 \rightarrow 60$) and several chair–chair interconversions can be observed. After 800 ps one of the chair conformations ($T_1 = 60$) exists almost exclusively. In a second study, three methyl-substituted cyclohexanes (1,1-dimethylcyclohexane, 1,1,3,3-tetramethylcyclohexane and 1,1,4,4-tetramethylcyclohexane) were subjected to molecular dynamics simulations at various temperatures. The observed chair–chair interconversions at the corresponding temperatures have been compared with experimentally determined energy barriers of ring inversion [28] (Table 1). As a result of the comparison it can be concluded that the molecular dynamics simulations are able to reflect the relative magnitude of the experimentally determined ring inversion barriers. This example of high temperature molecular dynamics clearly indicates the necessity to verify if the chosen simulation temperature is high enough to prevent the system from getting stuck in one particular region of conformational space.

Table 1. Data on the existence of the two possible chair conformations (chair and chair') of three methyl-substituted cyclohexanes at different simulation temperatures. The data are compared with the corresponding experimentally determined ring inversion barriers

Molecular form	Temperature				
	600 K	800 K	1000 K	1200 K	ΔG (kcal/mol^{-1})
	Chair	Chair + Chair'	Chair + Chair'	Chair + Chair'	9.6
	Chair	Chair	Chair + Chair'	Chair + Chair'	10.6
	Chair	Chair	Chair	Chair + Chair'	11.7

In the application of molecular dynamics to search conformational space it is a common strategy to select conformations at regular time intervals and minimize them to the associated local minimum. This procedure has been used in several conformational analysis studies on small molecules, including ring systems [14,29]. A very impressive example in this context is the conformational analysis of the polyhydroxy analog of the sesquiterpene lacton tharpsigargin (Fig. 6). This study was also performed in the laboratory of F. S. Jorgensen.

The polyhydroxy derivative has been studied in molecular dynamics simulations at 1200 K in order to gain insight into the conformational behavior of the ring system*. The seven-membered ring adopted several different conformations during the simulation and a

---

* Sybyl (version 6.0.3) from Tripos Associates Inc., St. Louis, U.S.A. Energy minimizations: Tripos force field, PM3 partial charges, dielectric constant $\varepsilon = 20$ and a convergence criterion of 0.005 kcal mol^{-1} Å^{-1}. MD simulations: 1000 ps at 1200 K with conservation of total energy, one conformation sampled per picosecond.

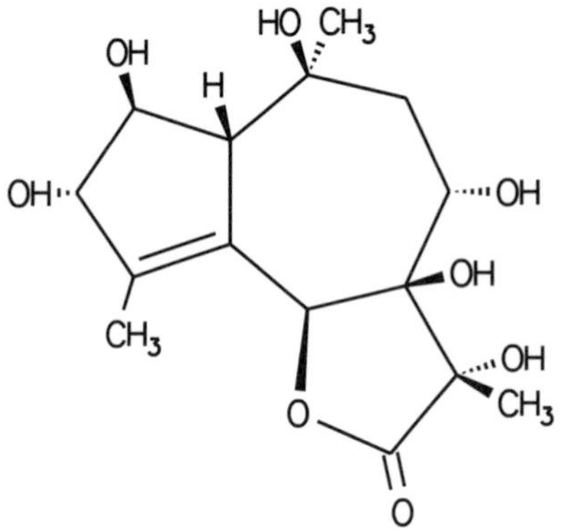

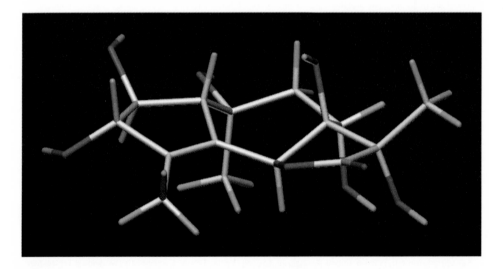

Figure 6. Molecular formula of the polyhydroxy analog of tharpsigargin.

considerable number of ring interconversions took place. This clearly demonstrates an extensive exploration of the conformational space.

Each of the sampled conformations has been energy-minimized subsequently and compared exclusively with respect to the conformation of the seven-membered ring. All conformations with a root mean square (rms) value below 0.1 Å were considered to be identical. The procedure yielded five different low-energy conformations. Fortunately NMR data [30] of tharpsigargin agree with one of the theoretically found conformations of the tricyclic ring system. This is shown in Fig. 7.

Figure 7. One of the theoretically determined conformations of the polyhydroxy analog of tharpsigargin. The ring conformation is in accordance with results obtained by NMR spectroscopy.

In some cases, however, it is not sufficient to minimize the sampled conformations in order to reach the final minimum conformation. The intention of the high-temperature dynamics simulation is to provide the molecule with enough kinetic energy to cross energy barriers between different conformations. However, during the simulation the molecule can occupy extremely distorted geometries which sometimes cannot be relaxed by a simple minimization procedure.

If this occurs it is recommended to perform a high-temperature annealed molecular dynamics simulation [31]. Using this approach all sampled conformations of the high-temperature simulation will be subsequently optimized and then reshaken at a lower temperature, e.g. 300 K, in order to remove the internal strain of the molecule. The final reoptimization leads to conformations of lower energy when compared with the results of a high-temperature simulation which is followed by a simple geometry optimization.

An additional modification of this high-temperature annealed molecular dynamics simulation is the so-called simulated annealing method [32]. In this technique the system is cooled down at regular time intervals by decreasing the simulation temperature. As the temperature approaches 0 K the molecule is trapped in the nearest local minimum conformation. The received geometry at the end of the annealing cycle is saved and subsequently used as starting point for further simulations at high temperature. In order to obtain a set of low-energy conformations the cycle will be repeated several times. As the resulting structures should already be close to a minimum it is not absolutely necessary subsequently to minimize the structure. The application of this method has been subject of several studies [33, 34]. Further information may be found in these references.

In conclusion, it may be stated that molecular dynamics simulations represent an additional and very valuable tool that can be used to sample the conformational space, especially when other conformational search methods have been unsuccessful. The user should be careful when selecting the appropriate method and in setting the simulation conditions in order to ensure the completeness of the conformational search and the validity of the results. It should also be kept in mind that each approach has its strengths and its weaknesses and therefore, wherever possible, experimentally derived data should serve as verification.

References

[1] Scheraga, H. A. *Chem. Rev.* **71**, 195–217 (1971).
[2] Rademacher, P. *Strukturen organischer Moleküle*. VCH: Weinheim 1987, p. 139.
[3] Ghose, A. K., Crippen, G. M., Revankar, G. R., Smee, D. F., McKernan, P. A., and Robins, R. K. *J. Med. Chem.* **32**, 746–756 (1989).
[4] Jörgensen, W. L. *Science* **254**, 954–963 (1991).
[5] Howard, A. E., and Kollman, P. A. *J. Med. Chem.* **31**, 1669–1675 (1988).
[6] Lipton, M., and Still, W. C. *J. Comput. Chem.* **9**, 343–355 (1988).
[7] Dammkoehler, R. A., Karasek, S. F., Shands, E. F. B., and Marshall, G. R. *J. Comput.-Aided Mol. Design* **3**, 3–21 (1989).
[8] Saunders, M. *J. Am. Chem. Soc.* **109**, 3150–3152 (1987).
[9] Saunders, M. *J. Comput. Chem.* **10**, 203–208 (1989).
[10] Ferguson, D. M, and Raber, D. J. *J. Am. Chem. Soc.* **111**, 4371–4378 (1989).
[11] Chang, G., Guida, W. C., and Still, W. C. *J. Am. Chem. Soc.* **111**, 4379–4386 (1989).

[12] Saunders, M., Houk, K. N., Wu, Y.-D., Still, W. C., Lipton, M., Chang, G., and Guida, W. C. *J. Am. Chem. Soc.* **112**, 1419–1427 (1990).

[13] Ghose, A. K., Jaeger, E. P., Kowalczyk, P. J., Peterson, M. L., and Treasurywala, A. M. *J. Comput. Chem.* **14**, 1050–1065 (1993).

[14] Böhm, H.-J., Klebe, G., Lorenz, T., Mietzner, T., and Siggel, L. *J. Comput. Chem* **11**, 1021–1028 (1990).

[15] Taylor, R., Mullier, G. W., and Sexton, G. J. *J. Mol. Graphics* **10**, 152–160 (1992).

[16] SYBYL Theory Manual, Tripos Associates, St. Louis, Missouri, USA.

[17] Maurhofer, E. *Konformationsuntersuchungen an calciumantagonistisch wirksamen Diphenylalkylaminen, Diphenylbutylpiperidinen, Phenylalkylaminen und Perhexilin*. Ph. D. Thesis, Freie Universität Berlin, Berlin, Germany 1989.

[18] Shenkin, P. S., and McDonald, D. Q. *J. Comput. Chem.* **15**, 899–916 (1994).

[19] Höltje, H.-D., and Batzenschlager, A. *J. Comput.-Aided Mol. Design* **4**, 391–402 (1990).

[20] Metropolis, N., Rosenbluth, A. W., Rosenbluth, M. N., Teller, A. H., and Teller, E. *J. Chem. Phys.* **32**, 1087–1092 (1953).

[21] Verlet, L. *Phys. Rev.* **159**, 98–103 (1967).

[22] Beeman, D. *J. Comp. Phys.* **20**, 130 (1976).

[23] Hockney, R. W., and Eastwood, J. W. *Computer Simulation Using Particles*. McGraw-Hill: New York 1981.

[24] van Gunsteren, W. F., and Berendsen, H. J. C. *Angew. Chemie* **102**, 1020–1055 (1990).

[25] Lybrand, T. P. Computer Simulation of Biomolecular Systems Using Molecular Dynamics and Free Energy Perturbation Methods. In: *Reviews in Computational Chemistry*, Vol. 1. Lipkowitz, K. B., and Boyd, D. B. (Eds.). VCH: New York; 295–320 (1990).

[26] McCammon, J. A., and Harvey, S. C. *Dynamics of Protein and Nucleic Acids*. Cambridge University Press: Cambridge 1987.

[27] Leach, R. A. A Survey of Methods for Searching the Conformational Space of Small and Medium-Sized Molecules. In: *Reviews in Computational Chemistry*, Vol. 2. Lipkowitz, K. B., and Boyd, D. B. (Eds.). VCH: New York; 1–47 (1991).

[28] Friebolin, H., Schmid, H. G., Kabuß, S., and Faißt, W. *Org. Magn. Reson.* **1**, 147–162 (1969).

[29] Kawai, T., Tomioka, N., Ichinose, Takeda, M., and Itai, A. *Chem. Pharm. Bull.* **42**, 1315–1321 (1994).

[30] Christensen, S. B., and Schaumburg, K. *J. Org. Chem.* **48**, 396–399 (1983).

[31] Auffinger, P., and Wipff, G. *J. Comput. Chem.* **11**, 19–31 (1990).

[32] Kirkpatrick, S., Gelatt, C. D., Vecchi, M. P. *Science* **220**, 671–680 (1983).

[33] Salvino, J. M., Seoane, P. R., and Dolle, R. E. *J. Comput. Chem.* **14**, 438–444 (1993).

[34] Laughton, C. A. *Protein Eng.* **7**, 235–241 (1994).

2.4 Determination of Molecular Interaction Potentials

The initial step in the formation of a complex like, for example, a drug–receptor complex is a recognition event. The receptor has to recognize whether an approaching molecule possesses the properties necessary for specific and tight binding. This recognition process occurs at rather large distances and precedes the formation of the final interaction complex. The 3D electrostatic field surrounding each molecule therefore plays a crucial role in recognition. Other molecular characteristics like polarizability or hydrophobicity come into play when the distance between the interacting surfaces gradually decreases. It is therefore easy to realize that molecular fields which can be determined by systematic calculation and sampling of interaction energies between the molecules under study using different chemical probes represent data sets of high value for the understanding of intermolecular interaction at any level of complexity of the molecular ensemble of interest.

In the following sections the methods for calculation and analysis of these molecular properties will be described and evaluated.

2.4.1 Molecular Electrostatic Potentials (MEPs)

Knowledge of the molecular electrostatic potential (MEP) is critically important when molecular interactions and chemical reactions are to be studied. If molecules approach each other, the initial contact arises from long-range electrostatic forces. In principle, interaction forces can be separated into three components: electrostatic, inductive and dispersive. The first type of interaction appears between polar molecules which carry a charge or possess a permanent dipole moment. The second type is found when a polar molecule interacts with a non-polar molecule. The dipole of the polar molecule then produces an electric field which changes the distribution of the electrons in the non-polar molecule, thereby inducing a dipole moment. Thirdly, even if both molecules are non-polar and hydrophobic entities, the permanent fluctuations in the electron distribution of one molecule can induce a temporary molecular dipole moment in a neighboring molecule. This type of interaction is called *dispersion*. Dispersion forces are weak and fall off rapidly with increasing distance between the interacting molecules (see section 2.2.1). However, they constitute the main part of attraction between neutral non-polar molecules. (The dispersion forces are also called London forces.)

The electrostatic interaction can be either attractive or repulsive; an electropositive portion of an approaching molecule will seek to dock with an electronegative region, while similarly charged portions will repel each other. The non-covalent interaction obviously is especially large between charged regions of molecules. Due to charges—but also due to permanent dipole moments present in a molecule—a 3D electrostatic field is generated in the surrounding environment. Therefore at moderate distances from polar or even neutral molecules, a significant molecular electrostatic potential exists. This can be represented as interaction energy between the molecular electron distribution and a positive point charge which is located in a 3D grid at any point in space surrounding the molecule. For the determination of the molecular electrostatic potential an accurate treatment of the electronic properties of the molecules is required. Therefore, methods for the calculation of molecular charge densities become priority.

2.4.1.1 Methods for Calculating Atomic Point Charges

The electronic properties of molecules are defined through the electron distributions around the positively charged nuclei. Detailed information about the electron distribution can be either obtained via experimental results, e.g. X-ray diffraction studies, or by calculations using quantum mechanical methods. However, with respect to the computational procedure corresponding results provide only a probability distribution of the charge density throughout three-dimensional space. For the purpose of interaction energy calculations mostly point charges located at the center of the atom positions are needed. Without doubt this produces a very simplified picture of the molecular electron distribution. To achieve the transformation the electron density needs to be converted into so-called partial or point charges. This can be done by contracting the charge onto the atomic centers. Thus, the picture of a molecule consisting of atoms carrying the partial or point charges has emerged. The definition of these empirical partial charges bears some arbitrariness because the molecular electron distribution must be assigned to individual atom centers. Or to put it in a different way, a molecular characteristic is scaled down to an atomic property. Partial charges are not observable, so the method of assigning point charges is only relevant and scientifically sound when it can be used to correlate or predict physical or chemical properties of molecules. On the other hand, as stated before, the electrostatic part of the overall intermolecular interaction energy is very prominent and therefore most of the commonly used molecular mechanics programs include a corresponding energy term which is dependent on atomic partial charges. The application of these methods allows the rapid computation of electrostatic energies, even for macromolecules with more than a few hundred atoms. For that reason a variety of different techniques for the calculation of atomic partial charges has been developed (for a review, see [1]).

In principle it must be distinguished between two methodologically absolutely different approaches:

1. Topological procedures [2–6] such as the Gasteiger–Hückel method [2].
2. Procedures which calculate atomic charges from the quantum chemical wave functions like the population analysis [7] or the potential-derived charge calculation methods [8–11].

Topological Charges

The topological methods are based mainly on the electronegativity of the different atom types. To allocate atomic charges to directly bonded atoms in a reasonable way, appropriate rules are used which combine the atomic electronegativities with experimental structural informations on the bonds linking the atoms of interest. The topological methods do not need information about the molecular geometry or conformational status of a molecule. Only the connectivity matrix of the atoms is included in the calculation. The original method proposed by Del Re [3] exclusively for saturated molecules was extended to conjugated systems by Berthod and Pullman [4]. Both methods still are implemented in some modeling programs. A newer approach, which gives more realistic results in comparison with experimental data is the Gasteiger-Hückel method. It is a combination of the Gasteiger–Marsili method [2] for the calculation of the σ component of the atomic charge and the old Hückel theory [12]. The Hückel theory allows to calculate the π component of the atomic charge in a fast and fairly efficient way. Naturally the total charge is the sum of σ and π elements. Formal charges on atoms included in π systems are assumed to be delocalized over the whole π system. For this reason, Hückel charges are calculated first and the Gasteiger charge calculation is performed subsequently. The big advantage of the topological procedures is that they are computationally fast and in many cases compare quite well with experimentally observable properties. The big danger is that one cannot trust the results without validation for a particular group of molecules. Very often the validation procedure simply is omitted. Of course this renders the corresponding study useless.

Topological methods often are implemented into commercial software packages as standard tools for charge calculation.

Quantum Chemical Methods

All other methods for the calculation of atomic partial charges are based on the quantum mechanical computation of wavefunctions. Wavefunctions either can be obtained using semiempirical or ab initio methods depending on the requested accuracy of the wavefunction and also on the available computational resources. Charge densities can be obtained from wavefunctions using different procedures. The oldest and most widely used is the Mulliken population analysis [7], which is implemented as standard method in various quantum mechanical programs [13–15]. The population analysis takes the electron density derived from the wavefunction and partitions it between the atoms on the basis of the occupancy of each atomic orbital. Although widely used, it has long been recognized in the literature that the results of the Mulliken method depends strongly on the basis sets applied. It often gives unrealistic results [16,17] (see also Table 1). An improved technique that eliminates most of the problems associated with the Mulliken procedure is the natural population analysis [18], but it is effective on ab initio wavefunctions only.

A second, much more recently developed, technique yielding atomic charges from quantum mechanically calculated wavefunctions is the method of deriving charges by fitting the molecular electrostatic potential (also called electrostatic potential (ESP) fit method) [7–11]. The charge density is a well-defined function [19]. It contains important and detailed information about the molecule because all electrons contribute in some way to the distribution of the electronic charge in space. It also is experimentally accessible [20] from X-ray diffraction. However, this technique is extremely demanding as far as costs and time consumption are concerned and cannot be used as a standard procedure. A set of atomic charges able to reproduce the 3D electron density seems to be an excellent choice for generating a fairly correct picture of the electronic properties of any molecule. The mathematical technique underlying the ESP fit method involves least-squares fitting of the atomic charges to reproduce as closely as possible the charge density, which has been calculated quantum mechanically at a set of points in space surrounding the molecule. This yields much better results [9,11] than the Mulliken population analysis.

Whether a charge distribution obtained with a particular method is reliable and able to represent realistically the electronic proportions of a molecule must be checked against

Table 1. Comparison of experimentally derived and theoretically calculated dipole moments. The theoretical dipole moments were calculated using several procedures: the Gasteiger–Hückel method was chosen as an example for simple topological methods; on the quantum mechanical level the dipole moments were calculated directly from the wavefunction (SCF) as well as using the Mulliken and potential-derived point charges (ESP)

Molecule	Experimental (gas phase)	Gasteiger-Hückel	AM1			PM3			STO-3G			3-21G*			6-31G**		
			SCF	Mulliken	ESP	SCF	Mulliken	ESP	SCF	Mulliken	ESP	SCF	Mulliken	ESP	SCF	Mulliken	ESP
Imidazole	3.8±0.4	3.118	3.508	2.129	3.575	3.861	2.412	3.869	3.535	2.213	3.494	4.025	2.855	3.962	3.855	2.822	3.810
Thiazole	1.61±0.03	1.466	2.012	2.680	2.041	1.249	1.463	1.259	1.986	2.554	1.989	1.683	3.556	1.709	1.435	2.594	1.507
Furane	0.66	0.599	0.493	0.354	0.484	0.216	0.066	0.234	0.532	0.675	0.498	1.101	2.222	1.087	0.772	1.813	0.738
Methyl-silane	0.735	–	0.374	0.276	0.331	0.432	0.175	0.402	–	–	–	0.702	1.572	0.238	0.672	0.027	0.658
NH_3	1.470	0.593	1.848	0.644	1.793	1.550	0.011	1.499	1.876	0.902	1.869	1.752	1.189	1.869	1.839	1.384	1.867
Dimethyl ether	1.31	1.764	1.429	1.052	1.473	1.254	0.854	1.3194	1.333	1.181	1.384	1.847	3.109	1.901	1.475	2.512	1.531

experimental data. One rather easy accessible experimental property is the molecular dipole moment. On the basis of atomic point charges a molecular dipole moment can be calculated in a simple and fast way and can be compared with appropriate experimental values which are listed for many compounds in literature (see for example [21]). Because the dipole moment depends crucially on the conformation of a molecule, only values for rigid molecules should be taken into consideration for comparative purposes. In order to decide on the applicability of a particular method for the calculation of charges in a series of molecules, one often proceeds by investigating not the entire flexible molecule but only small yet rigid fragments. Table 1 lists calculated and experimental dipole moments for a representative set of small and rigid structures. The dipole moments have been calculated using various methods and basis sets as well as the different procedures discussed earlier. The dipole moment is a quantum mechanically defined property; it can therefore also be calculated directly from wavefunctions (marked as SCF in Table 1). Corresponding results derived with a large basis set like 6-31G** are in especially good agreement with the experimental values.

Which type of procedure should be employed to investigate a particular molecular system depends on several factors. On the one hand the size of the molecules to be studied plays an important role; on the other the available computer power is the limiting factor for choosing a particular method.

Topological methods have the advantage over quantum chemical properties that they are very fast and give reasonable estimates of physical properties associated with charge. These methods generally produce dipole moments that are in good agreement with experimental values, partly a consequence of their calibration against experimental results. In contrast, the main disadvantage is the neglect of molecular geometries and conformations. Of course topological methods must fail in the case of molecules which contain atom types missing in the parameter list (see for example methylsilane in Table 1. Parameters for silicon are not included in the Gasteiger–Hückel method.).

Calculation of atomic charges from the molecular charge densities is the best choice if the results are for use in empirical energy functions for the purpose of interaction energy calculations. As can be deduced from Table 1 it is not absolutely necessary to use large ab initio basis sets. With smaller basis sets and even with the semi-empirical AM1 method dipole moments can be obtained which compare quite well with experimental values. However, the quality of the resulting dipole moment depends very distinctly on the procedure employed for generating the atomic point charges. All results obtained directly from molecular charge distribution are more realistic than the results of the Mulliken population analysis, which for some basis sets yields crude and erroneous dipole moments (see Table 1).

If a molecule of interest contains more than about 100 atoms then a sufficiently accurate calculation of the wavefunction is not feasible for the entire molecule. This impediment can be avoided by partitioning the large molecule into overlapping fragments. The fragment results then are transferred onto the large structure, hoping that the fragment properties correctly mirror the characteristics of the parent molecule.

However, even if point charges of high quality have been determined for a series of molecules these quantities are only weak arguments if the question of molecular similarity is the object of interest. Molecular similarity can be determined much more adequately on the basis of the 3D charge distribution. The most advantageous way to use this important and well-defined magnitude is through the MEPs.

2.4.1.2 Methods for Generating MEPs

MEPs are represented as interaction energies of a positively charged unit (a proton) with the charge density produced by the molecular set of nuclei and electrons at any point in space in the vicinity of the molecule. In general, a cut-off value is defined to limit the number of MEP points to be calculated. The MEP is a very useful tool in molecular modeling studies. It describes the electrostatic features of molecules and can be employed for the analysis and prediction of molecular interactions. For the generation of molecular electrostatic potentials two different approaches can be followed. The most desirable way is to calculate the MEPs directly from the quantum mechanically derived wavefunction. This procedure is straightforward and more accurate but time-consuming. A simpler approach is the calculation of MEPs on the basis of the atomic partial charges representing the molecular charge distribution. The MEP then is calculated applying the Coulomb equation for electrostatic interactions. Of course, the first procedure is by far superior and by all means should be used if sufficiently accurate wavefunctions are attainable for a particular molecule.

Many investigations are found in the literature which studied the basis set dependence of MEPs derived directly from wavefunctions [22–25]. It has also been shown that the electrostatic potential based on AM1 wavefunction correlates sufficiently well with ab initio results [22]. Therefore, AM1 can be used in all cases which cannot be handled due to molecular size at the ab initio level.

Visualization of MEPs

For the display of the molecular electrostatic potential different techniques are in use. The major obstacle for a fast and easy utilization of MEPs which permits the comparison of different molecules is the large amount of data points associated with this property. One very widely employed method to visualize MEPs is the display of the molecular electrostatic potential in the form of a 2D isocontour map in a particular plane of the molecule. The map may be displayed in color on a graphics screen, and can be manipulated in real-time. A single contour line represents values with similar energy. Regions containing a high nuclear contribution produce positive fields, corresponding to a repulsive interaction with a positive point charge, while those with a high electron density produce a negative potential, corresponding to an attractive interaction with a positive point charge.

The next level of complexity is reached by switching from 2D to 3D display mode. In principle, nothing changes since the molecule is completely wrapped by sets of isopotential shells. Each point on a particular shell experiences an electrostatic potential of the same sign and magnitude. With the help of this technique the overall distribution of positively and negatively charged regions around a molecule can be visualized very distinctly. While 2D charts naturally may not always reveal a complete picture of the molecular electrostatic potential, the 3D isopotential surfaces effectively allow qualitative interpretation and comparison to be made between different compounds.

The third method for displaying the molecular electrostatic potential is associated with the calculation and visualization of the molecular surfaces. We will therefore dwell only shortly on the various definitions of the molecular surface. In the formal treatment of molecular surfaces the atomic positions are treated as points, whereas the electron clouds are approximated by spheres centered on the atomic centers. If the electron spheres are represented by the van der Waals radii, then the surface generated by summing all spheres is called the *van der Waals surface*. Van der Waals surfaces approximately represent the 3D volume requirements of molecules. A different type of surface which is often used in molecular modeling studies is the *solvent accessible surface*, also called *Connolly surface* [26]. The Connolly surface is the surface encircled by the center of a solvent probe as the probe molecule rolls over the van der Waals surface.

The electrostatic potential can be color-coded either onto the van der Waals or the Connolly surface. Each color at a defined surface point on the surface indicates a distinct energy value of the electrostatic potential. This technique attempts simultaneously to display both the shape of the molecules as well as their electrostatic properties. However, when larger molecules are studied the images become very complex. A solution to this problem is sometimes found by using the different techniques in a combined approach, since areas hidden in one display mode may be perceptible in the other (see Fig. 1).

The molecular electrostatic potential is a much more reliable indicator of electrostatic reactivity than the concept of atomic point charges. MEPs and their 3D representation have proven to be effective tools for analyzing and predicting the interaction of ligands with their macromolecular receptors.

The electrostatic potentials of different molecules, which bind to the same receptor site in a similar way, must share common features. It has been shown that in many cases where an atom-by-atom fit of the corresponding molecules does not lead to a satisfactory result, the MEP-directed superimposition yields an acceptable solution of the problem (see section 2.5.3).

As an example, it has been shown in a study of the electrostatic potential of histaminergic H_2 antagonists [27] that the imidazole ring of cimetidine and the guanidinothiazole ring of tiotidine can be superimposed on the basis of their electrostatic potential. This can be easily deduced from Fig. 2.

2.4.2 Molecular Interaction Fields

Many biological processes are determined by non-covalent interactions between molecular structures. This is true for the docking of a ligand to a receptor, the interaction of a substrate with an enzyme, or the folding of a protein. Also in the world of crystals the non-covalent forces determine decisively the geometry and symmetry of the molecular arrangement. As a general rule binding only occurs if the generated energy of interaction overcomes the repulsive van der Waals forces. One method to investigate the energetic conditions between molecules approaching each other is the generation of molecular interaction fields. These fields describe

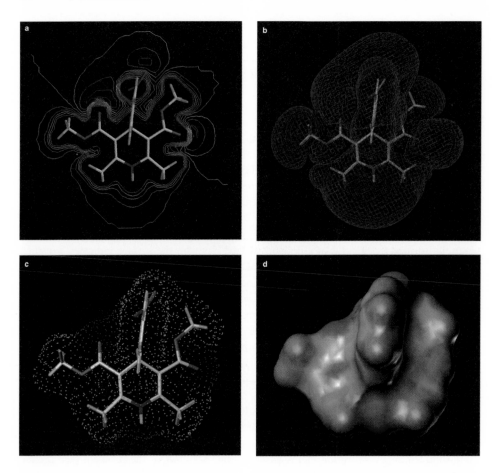

Figure 1. Visualization of the molecular electrostatic potential (MEP) of nifedipine, using different techniques. a) The MEP is displayed as a 2-dimensional isocontour map in the plane of the dihydropyridine ring system. The electrostatic potential has been calculated directly from the ab initio wavefunction (using a 6-31G** basis set) and is contoured from –50 kcal mol^{-1} (red) to 90 kcal mol^{-1} (blue). b) The MEP is displayed in the form of isopotential surfaces. The electrostatic potential has been calculated by a point charge approach (ESP point charges have been derived from an ab initio calculation applying a 6-31G** basis set) and is displayed at the region of –5 kcal mol^{-1} (blue) as well as 5 kcal mol^{-1} (red). (The calculations have been performed using the quantum mechanical software package SPARTAN 3.0 [14]). c, d) The electrostatic potential displayed on the Connolly surface of nifedipine. The values of the electrostatic potential have been calculated using ESP derived point charges [the same as in (b)] and are displayed in the form of a simple dot surface as well as in the more sophisticated form of a solid "triangular" surface. Blue areas represent negative electrostatic potentials; red areas represent positive values. (The calculations have been performed using program MOLCAD [58]).

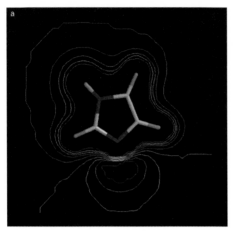 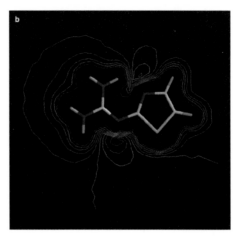

Figure 2. Electrostatic potential of imidazole (a) and guanidinothiazole (b). The electrostatic potentials have been calculated using the ab initio wavefunction (with a 6-31G** basis set) and are contoured from -50 kcal mol^{-1} (red) to 90 kcal mol^{-1} (blue). (The calculations are performed using the quantum mechanical software package SPARTAN 3.0 [14]).

the variation of interaction energy between a target molecule and a chemical probe moved in a 3D grid, which has been set around the target. The probes reflect the chemical characteristics of a binding partner, or fragments of it. By using computer graphics, molecular interaction fields can be displayed as 3D isoenergy contours. Contours of large positive energies indicate regions from which the probe would be repelled, while those of large negative energies correspond to energetically favorable binding regions.

The calculation of molecular interaction fields can be carried out using a variety of programs like GRID [28], CoMFA [29] or HINT [30]. GRID is one of the most widely used programs for investigating molecular interaction fields. It works for small molecules as well as large protein molecules such as enzymes. Only Cartesian coordinates are needed as input. The list of probes is very comprehensive and the interaction energy is calculated on a regular grid of points surrounding the target molecule. The grid can also be confined to a particular fragment of the target molecule if only this part is of interest. The calculated energies are stored in a datafile and can be transferred for graphical display and analysis into most of the common molecular modeling programs [31–33]. 3D contour maps may then be generated at any selected energy level and studied together with the target molecule on a computer graphics system. The contouring is a quick process which allows the user to control the graphical results almost immediately.

In this chapter we will focus on the calculation of the interaction fields for small molecules; investigations of the fields for macromolecules will be discussed later (see section 4.3.6).

2.4.2.1 Calculation of GRID Fields

The probes which can be used for the calculation are small molecules, chemical fragments or particular atoms, e.g. a water molecule, a hydroxyl group or a calcium ion. These probes simulate the chemical characteristics of the corresponding binding partners, for example a potential receptor protein binding site or the neighbor molecule in a crystal. In the course of a GRID calculation the probe is moved systematically through a regular 3D array of grid points around the target structure. At each point the interaction energy between the probe and the target is calculated using the following empirical energy function:

$$E_{tot} = E_{vdw} + E_{el} + E_{hb} \qquad (1)$$

where E_{tot} represents the total interaction energy, E_{vdw} represents the van der Waals interaction energy, E_{el} represents the electrostatic energy, and E_{hb} represents the interaction energy due to hydrogen bond formation.

The van der Waals interaction energy can be regarded as a combination of attractive and repulsive dispersion forces between non-bonded atoms. An atom of the probe is prevented from penetrating an atom in the target molecule by atomic repulsion and electron overlap. Repulsion forces can be estimated by an empirical energy function that becomes large and positive when the interatomic distance between two atoms is less than the sum of their van der Waals radii. The attractive part of the dispersion interaction is due to the correlated motion of electrons around the nuclei which results in induced dipole interactions. For non-polar molecules the balance between the attractive dispersion forces and the short-range repulsive forces can be described with the Lennard-Jones potential [34] (see for example Eq. (5) in section 2.2) which is implemented in the GRID program.

Electrostatic interactions are particularly important due to their long-range character for the attraction between ligand and macromolecular receptor.

The Coulomb equation (Eq. (6) in section 2.2) is widely used in molecular mechanics programs for the calculation of the electrostatic term because of its simple mathematical form. Its disadvantage is the fact that the heterogeneous media of molecular systems which consist of molecules with different dielectric properties are not sufficiently represented. The discontinuity between solute and solvent is taken into account by using an extended and more comprehensive form of Coulomb's law [28] which is used by GRID.

The directional properties of hydrogen bonds play a crucial role in determining the specifity of intermolecular interactions. It is therefore of utmost importance for a proper evaluation of interaction energies to describe this part of attractive forces between molecules in a correct form. A hydrogen bond can be regarded as intermediate-range interaction between a positively charged hydrogen atom and an electronegative acceptor atom [35]. The resulting distance between acceptor and donor atom is less than the sum of their van der Waals radii. In contrast to other non-covalent forces like dispersion and electrostatic point charge interactions, the hydrogen bonding interaction is directional, i.e. it depends on the propensity and orientation of the lone pairs of the acceptor heteroatom.

In order to comply with the requirements of these aspects the GRID method uses an explicit energy term for hydrogen bonds [36]. The functional form of this term has been developed to fit experimental data. All parameters are founded on experimental crystallographic data, i.e. direction, type and typical strength of these interactions are classified according to the real world of crystals.

The probes implemented in GRID are extensively defined by a variety of parameters, e.g. the hydrogen bonding possibility, the van der Waals radius or the atomic charge. The detailed description makes them very specific so that they can be regarded as realistic representatives of important functional groups found in the active site of macromolecules. As an example properties and parameters for three important probe groups are shown in Table 2.

Table 2. Examples of the different parameters which are necessary to define GRID probe groups

	Methyl probe	Hydroxyl probe	Carboxyl probe
Van der Waals radius (Å)	1.950	1.650	1.600
Effective number of electrons	8	7	6
Polarizabilty (Å^3)	2.170	1.200	2.140
Electrostatic charge	0.000	–0.100	–0.450
Optimal hydrogen bond energy (kcal mol^{-1})	0.000	–3.500	–3.500
Hydrogen bonding radius (Å)	–	1.400	1.400
Number of hydrogen bonds donated	0	1	0
Number of hydrogen bonds accepted	0	2	2
Hydrogen bonding type	0	4	8

GRID also contains a table of parameters defining each type of atom that possibly exists in target molecules. The respective parameters define the strength of the van der Waals, the electrostatic and the hydrogen bond interactions potentially formed by an atom. The careful parametrization and the great variety of implemented probes make the GRID program a precise and widely used method for the investigation of interaction fields for small molecules as well as for macromolecular structures.

2.4.2.2 How GRID Fields can be Exploited

The calculation of molecular interaction fields has been applied to a wide range of molecular modeling studies [37–42]. The strategy employed depends on the available structural information for ligands and macromolecular targets. If the 3D structure of a macromolecule is known, the interaction fields can be used to locate precisely favorable binding regions for the ligands. Subsequently these regions can be taken as a starting point for the design of new ligands for the particular receptor. This procedure will be described in detail in Chapter 3.

More often, situations are met where there is no structural information about the macromolecular receptor and only the properties of the ligands are available. Under such circumstances molecular interaction fields can help to generate a more or less detailed picture of the potential receptor binding site. A prerequisite for this approach of course is that all investigated ligand molecules indeed bind to the same receptor site in an analogous mechanism. Only then can they be expected to exhibit a similar interaction pattern. Also, relative positions and size of the contours at any given energy level should be comparable. The energy level at which the contours must be compared is highly dependent on the probe type chosen.

Two different types of interaction fields for the Ca^{2+}-channel-blocking agent nifedipine are shown as an example in Fig. 3.

The interaction fields mark those parts of the corresponding binding region which possess particular chemical and physical properties. These properties can be translated into a model of the binding region of the macromolecule. If the macromolecular target is a receptor protein the model is composed of single amino acid fragments which are located at the corresponding interaction regions. The amino acid fragments should satisfy the different binding regions which are common for all the active compounds. For example, the hydrophobic interaction fields possibly represent the location of hydrophobic amino acids such as phenylalanine, tryptophan, valine, leucine or isoleucine. Of course further investigations are necessary to specify the exact type of amino acid in each case. This will be discussed in the next section where this approach, called *receptor mapping,* will be described in detail.

If a large set of compounds has to be studied it may become difficult to recognize all existing common interaction patterns. One way to solve this problem is to calculate the common interaction regions for different target molecules which were obtained in each case using the same probe. The common regions are mathematically detected in a gridpoint-by-gridpoint comparison of the fields. Only the hits are saved in a file and used for the generation of a common interaction field [43].

A more profound technique for comparison and analysis of molecular interaction fields is the use of chemometric methods like GOLPE [44] or PLS [45]. Until recently most structure–activity relationship studies based on molecular fields have been of a qualitative nature. One probable reason for this is that the methods for statistical evaluation contain mathematical and methodical difficulties that make these methods practicable only for specialists in chemometrics. Nevertheless, qualitative analyses have also shown in many studies

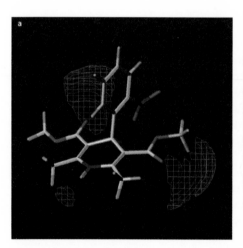 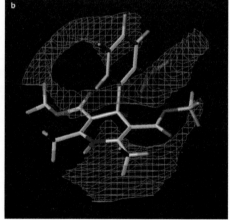

Figure 3. Visualization of the molecular interaction fields of nifedipine. a) Favorable hydrogen-bonding regions derived from GRID [28] calculations using a hydroxyl probe (contour level: –3.5 kcal mol^{-1}). The favorable hydrophobic interaction regions obtained with a methyl probe are displayed in b) (contour level: –1.4 kcal mol^{-1}).

[36–43] the value of molecular interaction fields in identifying features contributing to bioactivity.

2.4.2.3 Use of Chemometrics: The CoMFA Method

The CoMFA [29] method (Comparative Molecular Field analysis) was developed as a tool to investigate 3D quantitative structure–activity relationships (3D-QSARs). 3D-QSAR approaches use statistical methods (chemometrical methods) to correlate the variation in biological or chemical activity with information on the 3D structure for a series of compounds. The underlying idea of the comparative molecular field analysis is that differences in a target property, e.g. biological activity, are often closely related to equivalent changes in the shapes and strenghts of the non-covalent interaction fields surrounding the molecules. Or stated in a different way, the steric and electrostatic fields provide all the information necessary for understanding the biological properties for a set of compounds. As in the GRID approach, the molecules are located in a cubic grid and the interaction energies between the molecule and a probe are calculated for each grid point. A very important prerequisite for this procedure is some sort of alignment for the set of molecules. The alignment can be achieved by the well-known method of pharmacophore determination (see Chapter 3). According to the authors of CoMFA the method itself should also be used for this purpose. For this reason a typical CoMFA study starts with a rough alignment of the compounds. After calculation and sampling of the interaction energies at all predefined gridpoints (which are located at the intersections of a 3D lattice) the molecules so to speak are represented by their steric and electrostatic field properties. The relative 3D positions of common regions in the fields can be discovered with the help of statistical and chemometrical methods like for example PLS (for detailed information see [45]). The discovered common regions of field properties can be used subsequently to optimize the superimposition of the test structures in a so-called "Field Fit" procedure. This means that in cases where active molecules belong to different structural families, and cannot be aligned using an atom-by-atom fit procedure, they might be superimposable on the basis of common molecular interaction fields.

Although this method is of general use a word of caution is necessary. There are a number of practical problems that emerge in the course of its application. The results depend critically on the chosen ligand conformation, on the reasonableness of the alignment, on the chemical parameters used to describe the interaction fields, and last but not least on the selected statistical evaluation method [46]. The reader should be aware of the fact that this program is a powerful tool in the hand of the experienced user but may provide some difficulties for beginners. For that reason only a short description of the CoMFA method has been given here. A detailed description of all features and difficulties that are related to CoMFA would be beyond the scope of this chapter. (For a detailed description of the different CoMFA approaches, see [47].)

2.4.3 Hydrophobic Interactions

As we have already discussed, attraction and repulsion between molecules are controlled by various types of interaction. One type which has not yet been considered is the so-called hydrophobic interaction. Hydrophobic interaction between molecules is a complex process that results primarily from entropic effects related to the change in the orientation of solvent molecules in the solvation shell wrapping the solute molecules, but also from the bulk form of the solvent molecules. For an effective hydrophobic interaction a close contact of the interacting hydrophobic surfaces is necessary [48,49]. The following piece of fiction might lead to a better understanding of hydrophobic bonding: a non-polar region of a deep binding cavity in a protein is not directly solvated. Nearby water molecules shield the cavity and are thought to form an iceberg-like structure which is stabilized by intermolecular hydrogen bonds between the water molecules. An interaction between the hydrophobic surfaces of the binding cavity and an entering substrate leads to a disruption of the ordered iceberg-like structure. The disruption yields an increase in entropy which results in a gain of free energy for the total system [49]. The desolvation of the substrate molecule also of course adds to the amount of newly formed bulk solvent molecules and must be taken into account. To date, the entropic effect usually is ignored in most modeling studies, because no simple method of calculation is available. On the other hand, it is generally accepted that the hydrophobic bonding or entropic effect does indeed play an important role in each drug–receptor interaction [50] as well as protein folding [51] event. As a natural follow-up the hydrophobic interaction should by all means be included in the energy balance of these processes.

2.4.3.1 Log *P* as a Measure of Lipophilicity

Hydrophobicity can also be regarded as an empirical property of molecules encoding specific thermodynamic information about a molecule's interaction with its environment. Hitherto, several attempts were made for taking into account hydrophobic effects on the basis of experimental findings. The most important experimental measure of hydrophobicity is the solvent partition coefficient—expressed as log *P*—of a molecule between water and an organic phase. Since the log *P* can be determined experimentally it is a very useful tool. It can also be used to control and improve empirically developed methods, which are reported by several authors [52–54]. The prediction of log *P* can be achieved by transforming experimental solvent partition data for sets of variously substituted molecules into so-called hydrophobic fragment constants. These fragment constants represent the relative lipophilicity of particular structural elements found in the original set of molecules. The total lipophilicity of a compound (given by the log *P*) then can be calculated by summation of all fragment constants for a molecule under study. Today, fragment constants are available for a great variety of organic species with biological importance.

It should be noted that log *P* is a simple "one-dimensional" representation of hydrophobicity and only reflects an overall property. It is insufficient if a more detailed insight into molecular interactions between ligands and macromolecules is needed.

2.4.3.2 The Hydropathic Field

For that reason several attempts were made to utilize solvent partition coefficients as foundation to create a 3D representation for hydrophobicity. One way of approaching the problem is the generation of a hydrophobic field in analogy to the electrostatic field. This technique for example is implemented in the program HINT [30].

The HINT model of hydrophobic interactions is based on the fact that solubility data can be regarded as just another physical property capable of mirroring the molecular interactions between solute and solvent molecules. In the framework of the HINT theory the fragment-level solvent partition data (the hydrophobic fragment constants) are reduced to hydrophobic atom constants [55]. These atomic descriptors are the key parameters and must be assigned for each atom in a molecule under investigation. Since the hydrophobic atom constants are derived from experimental partition experiments and solubility data, the obtained hydrophobic atom constants not only model the hydrophobic interactions but also include other types of molecular interactions, like electrostatic and van der Waals terms. The generated field therefore incorporates hydrophobic as well as hydrophilic parameters. It is called a *hydropathic field*. The calculation is performed using an empirical function (for a detailed description of the functional form, see [55]). The hydrophobic atom constants, the section of the solvent accessible surface created by each atom, and a distance function are included in the algorithm. The distance function is necessary to describe adequately the distance dependence of the hydrophobic effect in the biological environment. HINT generates 3D molecular grid maps in a similar way as discussed for comparable programs.

The result of a HINT study is a combined contour map for the hydrophobic and hydrophilic field around a molecule. Grid points with a positive sign represent a hydrophobic region. The opposite is true for hydrophilic (polar) segments of space. Because of the empirical nature of the data, it is difficult to decide at which energy niveau the fields have to be contoured. It is self-evident that the selected energy level directly determines the size of the visualized part of the field. For a proportionally correct balance in the size of the displayed contours it is usually advisable to contour the hydrophilic effect at a level 2–5 times higher than that of the hydrophobic effect [56].

The appearance of hydrophobic and hydrophilic fields again is demonstrated for the well-known Ca^{2+} channel blocker, nifedipine in Fig. 4(a).

The information obtained from the analysis of the hydropathic field can be exploited following different strategies. The qualitative information on the distribution of hydrophobic and polar properties in the vicinity of a series of molecules for example can be used to generate a 3D map of the unknown receptor macromolecule. If the investigated series is large and complex an interface allows the produced data set to be read directly into CoMFA for a more elaborate analysis [57].

If the structure of the macromolecular receptor is known, the generated hydropathic fields also can be used to optimize the structures of ligands for enhancement of the biological activity. For a review of other potential applications, see [57].

2.4.3.3 Display of Properties on a Molecular Surface

The display of hydrophobic and hydrophilic property distributions in the extramolecular space can also be projected onto a molecular surface. The program MOLCAD [58] employs for example the Connolly surface [26] of a molecule as a screen for mapping local molecular properties such as lipophilicity by a color-coded representation. A distance-dependent

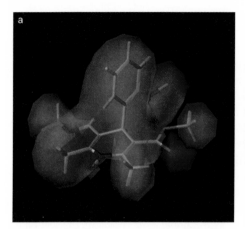 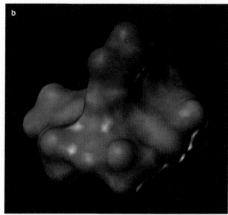

Figure 4. a) Hydropathic field map of nifedipine. The green surface represents the hydrophobic area and the red surface the hydrophilic area of nifedipine. The map has been calculated by HINT 2.02 [30] and is contoured at –8 (red) and 4 (green). b) Molecular lipophilic potential of nifedipine displayed on the Connolly surface. Brown areas on the surface represent more lipophilic parts and blue areas the hydrophilic parts of the molecule. (The calculation has been performed using program MOLCAD [32]).

function must be defined in order to reflect correctly the influence of different atoms or fragments on the local lipophilicity at a certain point on the molecular surface. This can be realized for example by introducing a molecular lipophilicity potential [59], which can be regarded as a pendant to the molecular electrostatic potential. As in the case of the MEP, the projection of any local properties onto a surface facilitates the perception and interpretation of the distribution of the visualized property descriptor. The main advantage of the surface-bound representation of hydrophobicity is the fact that the analysis of large molecular systems, like proteins, is much easier in comparison with the evaluation of hydropathic fields. Because the theoretical background of both methods is equivalent, the results obtained should be comparable qualitatively. For both methods an effective test of reliability can be performed for all molecules for which experimentally derived log P values are available. However, the partition coefficient—like the charge distribution—is drastically influenced by the conformation of a molecule. Moreover, the situation is further complicated by the conformation of a molecule being able to change when it migrates from the aqueous to the lipophilic environment. Unfortunately this fact limits the amount of test molecules to a rather small collection of rigid or at least semirigid structures. An example for the appearance of MOLCAD hydrophobic surfaces is shown in Fig. 4(b).

References

[1] Williams, D. E. Net Atomic Charge and Multipole Models for the ab Initio Molecular Electric Potential. In: *Reviews in Computational Chemistry*, Vol. 4. Lipkowitz, K. B., and Boyd, D. B. (Eds.). VCH: New York; 219–271 (1991).

[2] Gasteiger, J., and Marsili, M. *Tetrahedron* **36**, 3219–3228 (1980).

[3] Del Re, G. A. *J. Chem. Soc. London*, 4031–4040 (1958).

[4] Berthod, H., and Pullman, A. *J. Chem. Phys.* **62**, 942–946 (1965).

[5] Abraham, R. J., and Hudson, B. *J. Comput. Chem.* **6**, 173–185 (1985).

[6] Mullay, J. *J. Am. Chem. Soc.* **108**, 1770–1779 (1986).

[7] Mulliken, R. S. *J. Chem. Phys.* **23**, 1833–1840 (1955).

[8] Momany, F. A. *J. Chem. Phys.* **82**, 592 (1978).

[9] Cox, S. R., and Williams, D. E. *J. Comput. Chem.* **2**, 304–323 (1981).

[10] Singh, U. C., and Kollman, P. A. *J. Comput. Chem.* **5**, 129–145 (1984).

[11] Chirlian, L. E., and Francl, M. M. *J. Comput. Chem.* **8**, 894–905 (1987).

[12] Purcel, W. P., and Singer, J. A. *J. Chem. Eng. Data* **12**, 235–246 (1967).

[13] Frisch, M., Binkley, J. S., Schlegel, H. B., Raghavachari, K., Martin, R., Stewart, J. P., Bobrowicz, F., Defrees, D., Seeger, R., Whiteside, R., Fox, D., Fluder, E., and Pople, J. A. GAUSSIAN. Department of Chemistry, Carnegie Mellon University, Pittsburgh, Pennsylvania.

[14] SPARTAN, Wavefunction Inc., Irvine, California, USA.

[15] Schmidt, M. W., Boatz, J. A., Baldridge, K. K., Koseki, S., Gordon, M. S., Elbert, S. T., and Lam, B., GAMESS, Program No. 115, Quantum Chemistry Program Exchange, Indiana University, Bloomington, Indiana.

[16] Williams, D. E., and Yan, J. M. *Adv. Atomic Mol. Phys.* **23**, 87 (1988).

[17] Wiberg, K. E., and Rablen, P. R. *J. Comput. Chem.* **14**, 1504–1518 (1993).

[18] Reed, A. E., Weinstock, R. B., and Weinhold, F. A. *J. Chem. Phys.* **83**, 735–743 (1985).

[19] McWeeney, R. *Methods of Molecular Quantum Mechanics*, 2nd Ed. Academic Press: San Diego 1989.

[20] Destro, R., Bianchi, R., and Morosi, G. *J. Phys. Chem.* **93**, 4447–4457 (1989). Destro, R., Bianchi, R., Gatti, C., and Merati, F. *Chem. Phys. Letters* **186**, 47–52 (1991).

[21] McClellan, A. L. *Tables of Experimental Dipole Moments*, Vol. 2, Rahara Enterprise: El Cerrito, Californien, USA.

[22] Ferenczy, G. G., Reynolds, C. A., and Richards, W. G. *J. Comput. Chem.* **11**, 159–169 (1990).

[23] Rodriguez, J., Manaut, F., and Sanz, F. *J. Comput. Chem.* **14**, 922–927 (1993).

[24] Ford, G. P., and Wang, B. *J. Comput. Chem.* **14**, 1101–1111 (1993).

[25] Price, S. L., Harrison, R. J., and Guest, M. F. *J. Comput. Chem.* **10**, 552–567 (1989).

[26] Connolly, M. L. *Science* **221**, 709–713 (1983).

[27] Höltje, H.-D., and Batzenschlager, A. *J. Comput.-Aided Mol. Design* **4**, 391–402 (1990).

[28] Goodford, P. J. *J. Med. Chem.* **28**, 849–857 (1985).

[29] Cramer, R. C., Patterson, D. E., and Bunce, J. D. *J. Am. Chem. Soc.* **110**, 5959–5967 (1988).

[30] Kellogg, G. E., Semus, S. F., and Abraham, D. J. *J. Comput.-Aided Mol. Design* **5**, 545–552 (1991).

[31] INSIGHT/DISCOVER, Biosym Technologies Inc., San Diego, California, USA.

[32] SYBYL, Tripos Associates, St. Louis, Missouri, USA.

[33] MACROMODEL, Mohamadi, F., Richards, N. G. C., Guida, W. C., Liskamp, R., Carfield, C., Chang, G., Hendrickson, T., and Still, W. C. *J. Comput. Chem* **11**, 440–464 (1990).

[34] Lennard-Jones, J. E. *Proc. Roy. Soc.* **106A**, 463–477 (1924).

[35] Dean, P. M. *Molecular Foundations of Drug–Receptor Interaction*. Cambridge University Press: Cambrige 1986.

[36] Wade, R. C. Molecular interaction fields. In: *3D QSAR in Drug Design – Theory, Methods and Applications*, Kubinyi, H. (Ed.). ESCOM Science Publishers B. V.: Leiden; 486–505 (1993).

[37] Wade, R. C., Clark, K. J., and Goodford, P. J. *J. Med. Chem.* **36**, 140–147 (1993).

[38] Reynolds, C. A., Wade, R. C., and Goodford, P. J. *J. Mol. Graphics* **7**, 103–108 (1989).

[39] Meng, E. C., Shoichet, B. K., and Kuntz, I. D. *J. Comput. Chem.* **13**, 505–524 (1992).

[40] Höltje, H.-D., and Jendretzki, U. *Pharm. Pharmacol. Lett.* **1**, 89–92 (1992).

[41] Wade, R. C., and Goodford, P. J. *Br. J. Pharmacol. Proc. Suppl.* **95,** 588 (1988).
[42] Cruciani, G., and Watson, K. A. *J. Med. Chem.* **37**, 2589–2601 (1994).
[43] Höltje, H.-D., and Anzali, S. *Die Pharmazie* **47**, 691–697 (1992).
[44] Baroni, M., Costantino, G., Cruciani, G., Riganelli, D., Valigi, R., and Clementi, S. *Quant. Struct.-Act. Relat.* **12**, 9–20 (1993).
[45] Wold, S., Johansson, E., Cocchi, M. PLS – Partial Least-Squares Projections to Latent Structures. In: *3D QSAR in Drug Design – Theory. Methods and Applications.* Kubinyi, H. (Ed.). ESCOM Science Publishers B. V.: Leiden; 523–550 (1993).
[46] Klebe, G., and Abraham, U. *J. Med. Chem.* **36**, 70–80 (1993).
[47] Folkers, G., Merz, A., and Rognan, D. CoMFA: Scope and Limitations. In: *3D-QSAR in Drug Design – Theory, Methods and Applications.* Kubinyi, H. (Ed.). ESCOM Science Publishers B.V.: Leiden; 583–618 (1993).
[48] Tandford, C. *Science* **200**, 1012–1018 (1978).
[49] Tandford, C. *The Hydrophobic Effect.* Wiley: New York 1980.
[50] Suzuki, T., and Kudo, Y. *J. Comput.-Aided Mol. Design* **4**, 155–198 (1990).
[51] Nicholls, A., Sharp, K. A., and Honig, B. *Proteins* **11**, 281–296 (1991).
[52] Hansch, C., and Fujita, T. *J. Am. Chem. Soc.* **86**, 1616–1626 (1964).
[53] Rekker, R. F., and Mannhold, R. *Calculation of Drug Lipophilicity.* VCH: Weinheim 1992.
[54] Ghose, A. K., and Crippen, G. M. *J. Comput. Chem.* **7**, 565–577 (1986).
[55] Kellogg, G. E., Joshi G. S., and Abraham, D. *J. Med. Chem. Res.* **1**, 444–453 (1992).
[56] Kellogg, G. E., and Abraham, D. *J. Mol. Graphics* **10**, 212–217 (1992).
[57] Abraham, D. J., and Kellogg, G. E. Hydrophobic Fields. In: *3D QSAR in Drug Design – Theory, Methods and Applications.* Kubinyi, H. (Ed.). ESCOM Science Publishers B. V.: Leiden; 506–522 (1993).
[58] Heiden, W., Moeckel, G., and Brickmann, J. *J. Comput.-Aided Mol. Design* **7**, 503–514 (1993).
[59] Furet, P., Sele, A., and Cohen, N. C. *J. Mol. Graphics* **6**, 182–189 (1988).

2.5 Pharmacophore Identification

2.5.1 Molecules to be Matched

In the first sections of this book we have described how physico-chemical characteristics of molecules can be calculated and visualized. Now, we will discuss how this knowledge can be used to understand or predict the pharmacological properties of a compound. In the large majority of cases the basis for a pharmacodynamic effect is the interaction of a certain substance with a protein of physiological importance. The macromolecule might be an enzyme or a receptor. In both cases there must exist a highly specific 3D cavity which serves as binding site for the drug molecule. Compounds exerting qualitatively similar activities at the same enzyme or receptor therefore must possess closely related binding properties. That is, these molecules must present to the macromolecular binding partner structural elements of identical chemical functionality in sterically consistent locations. In short, congeners of a defined pharmacological group possess an identical pharmacophore, and one of the major tasks to be solved using molecular modeling techniques is the determination of pharmacophores for congeneric groups of drug molecules. Because the 3D structures of most receptors hitherto remain undiscovered, information on the corresponding hypothetical pharmacophore as a matter of fact is a very important source for understanding drug–receptor interactions at the molecular level.

When all physico-chemical properties have been intensively studied the question remaining is "How do we have to superimpose the members of a series to find the pharmacophore?" In order to answer this question we have first to define the pharmacophoric elements. That is, we must decide what functional groups or atoms have to be superimposed. Of course this question cannot be answered completely objectively in an automatic procedure because one always has to decide in advance on the atom pairs which correspond between two molecules. This may produce a large number of useless data if known structure–activity relationship information is not included. This facilitates the superpositioning procedure, because it drastically limits the number of solutions. It should be noted that similarity between ligands must not comprise the whole molecule, because most of the ligand molecules are not completely wrapped by receptor binding sites when they are bound to it. This also reduces the number of reasonable solutions.

If hydrogen bonds are supposed to be important for the pharmacophore then the direction and distance of lone pairs should be added to the atomic pattern of the molecules under study. This can be realized for example by locating dummy atoms in corresponding positions. These positions then are labelled by different flags as hydrogen bond-acceptor or -donor sites (only hydrogens bound to heteroatoms) and can be used as a first test for a superposition mode (program AUTOFIT [1]). Furthermore, planar elements like aromatic ring systems can be treated as special structural units. In this case for example the center of the ring system can be defined as matching point instead of the ring system. Other planar groups can be handled analogously.

If the set of molecules contains only very flexible congeners then the search for a common pharmacophore is not only very difficult and tedious but also might even yield either none or an arbitrary (and therefore useless) result. This task can be easily performed and is of far greater significance if rigid or at least semi-rigid compounds are present. These of course must be highly active, otherwise they can not be used as matrices for the flexible ligands. By the same token, the consideration of highly active but conformationally restricted molecules relieves the need to prove that one is indeed dealing with bioactive conformations.

The selection of the molecules to be superimposed is very important if significant results are to be obtained. The easiest to perform, but rather ineffective, case is the superimposition of structurally very similar compounds. This does not provide much information, so it is much more effective to include in the series structures containing different skeletons. As a natural follow-up this leads to a situation which is highly desirable where a simple atom-to-atom superpositioning is not possible but a matching of functionally equivalent elements or a matching of molecular fields must be performed.

One further point must be addressed. Are inactive molecules or molecules with only low activity to be taken into consideration? It seems useful at first to superpose highly active molecules alone. The derived pharmacophore then can be tested against, and eventually modified by, inclusion of low active and inactive congeners. The same holds true for antagonists and agonists of one receptor type. Superpositions should be performed for both groups separately. However, both models subsequently can possibly be combined, because very often competitive antagonists are bound at least partially in the agonistic receptor binding site. However, it should be noted that overlapping binding sites of agonists and antagonists are indeed common but do not necessarily exist.

Several different superpositioning procedures are available. They comprise manual or automatic fitting by rigid-body rotation or flexible-fitting procedures where both root mean square (rms) derivation between the fitted atom pairs and conformational energies are minimized. Other important superpositioning techniques perform alignments on the basis of equivalences detected in molecular surfaces or molecular field properties.

2.5.2 Atom-by-Atom Superposition

The least-squares technique for superpositioning of corresponding atom positions is the most widely used method. Two molecules are superimposed by minimizing the rms of the distances between the corresponding atom pairs in the molecules. The rms value represents a measure for the quality of the fit. This procedure is very powerful in discovering dissimilarity between molecules which seem to be apparently similar. The weak point is that it is required to decide in advance which atom pairs match. It is obvious that different superpositions are obtained depending on the atoms used for the procedure. The method cannot be applied to molecular systems in which atom-to-atom correspondences are not detectable in advance. However, rigorous similarity in atomical structure is not a prerequisite for the interaction of different molecules with the same receptor. Therefore for a large number of cases where pharmacological data and structure–activity studies urge upon a common mechanism of action for a set of dissimilar molecules the conventional least-squares superpositioning method is considered inadequate.

One may try to escape such a situation by performing a manual, interactive superposition if the test set is not too large. In principle, any number of molecules can be investigated directly on the graphics display and the fit may be judged visually. Certainly this procedure is very creative and may stimulate new ideas about the underlying mechnism of experimentally detected structure–activity data. On the other hand, such a procedure naturally is biased and often cannot correctly be reproduced, because a computational optimization is not applicable.

An efficient and fast search technique which can be used very successfully for the generation of pharmacophore models, the Active Analogue Approach, was developed by Marshall et al. [2, 3]. This technique utilizes a systematic search algorithm for calculating a representative number of sterically and energetically allowed conformations for congeneric molecules. For each of these conformations a set of distances between pharmacophoric groups believed to be important for the interaction with the receptor is generated. If each set of distances for one molecule is compared with all sets of all the other molecules—with the intention to find possible correspondences—the problem would not be solvable except for small molecules. On the other hand, in the framework of pharmacophore identification the complete conformational space of all compounds is not of interest but rather only those subregions which are accessible to all active ligands. As we have discussed earlier, it is of major advantage to include rigid or semi-rigid compounds in a conformational analysis for a series of flexible molecules. For that reason the conformational search is started with the most rigid molecule. After determination of the respective distance map for this compound these distances are used as constraints in the conformational search runs for the more flexible molecules. Following these lines the results of a search on one active and rigid analog are taken as a basis to explore the conformational space of all the other congeners of the series. As all of the active compounds must fit the receptor model the search is restricted to those regions of conformational space which correspond to the previously defined model. For example, according to the model if a pair of atoms must lie within a certain distance range in order to agree with the constraints, then only those torsions that will allow this constraint to be satisfied need to be calculated. An example which has demonstrated the strength of the Active Analogue Approach dealt with 28 angiotensin-converting enzyme (ACE) inhibitors in an effort to predict a model for the ACE active site [4]. Applying this technique the search time was reduced by more than three orders of magnitude in comparison with a previously performed conventional systematic search study on the same subject.

Another mapping procedure, which in contrast does not use an explicit atom-by-atom superposition approach is SEAL [5]. This program allows a rapid pairwise comparison of dissimilar molecules. The similarity score, as an indicator of the quality of fit, is calculated from a summation over all possible atom pairs between the two molecules. Each atom pair is weighted by the relative distance between the contributing atoms. In doing so the alignment function considers all theoretically possible atom pairs in the molecules in the comparison procedure and not only one atom pair, as in the atom-by-atom fit approach. As a consequence the resulting superposition reflects to some extent the properties associated with the global shape of the molecules. The program also offers the possibility to include physico-chemical properties in the alignment procedure. Therefore, the terms used in the pairwise summation can be composed from any physico-chemical quality supposed to be important for the biological effect. In the original version the authors used only van der Waals radii as an expression of sterical volume as well as point charges mimicking the electronic molecular properties to optimize the alignment.

There also exist mapping techniques which include as one of the first steps in the computational protocol the automatic and therefore unbiased identification of atomic centers or site points as correspondences used for superpositioning. Site points may include points of the molecular surface representing molecular features like hydrogen bond acceptor or donor characteristics. Several commercial program packages like APEX-3D [6] and CATALYST [7] offer such functionalities. Others like DISCO [8], RECEPS [9], and AUTOFIT [1] have been discussed in the literature. As described earlier the superposition is performed by matching the assigned corresponding atoms or site points in all possible combinations. The ranking of the achieved alignments is done by rms calculations.

2.5.3 Superposition of Molecular Fields

Since molecules recognize each other by characteristic properties on or outside their van der Waals volume—and not through their atomic skeleton—the determination of molecular similarity should take into account the molecular fields. As a natural follow-up the superpositioning approaches should also concentrate on mapping and comparing these properties. For matching purposes the molecules are located in a 3D grid of equally spaced field points. Each grid point is loaded with a certain characteristic property measure such as charge distribution, hydrophobic potential or simply information on the size of the volume. Similarity thresholds can be defined in order to guide the optimization procedure to a significant and unequivocal result. Single grid points or clusters of adherent grid points can be assigned different weights in order to pay as close as possible attention to structure–activity relationships. One molecule—preferentially with limited conformational freedom—is chosen as the template molecule. The grid loadings of the template serve as a measure for the various properties and all trial molecule grids are manipulated by rotation and translation to find the best fit of the grid values. The computational technique of orientational search which has to be used is extremely time-consuming. Different methods have been described which mirror different levels of complexity but also utilize various field properties. Manaut et al. [10] reported an effective method which maximizes the similarity between molecular surfaces on the basis of the molecular electrostatic field. Other groups such as Clark et al. [11] or Dean et al. [12] use physico-chemical field properties calculated using Lennard-Jones potentials, or replace the regular grid-based evaluation technique by an integration over Gaussian-type functions to approximate the electrostatic potential. Goodness-of-fit indices can be calculated for example as ratio of the number of commonly occupied grid points to the total number of grid points.

In summary, the tools for matching molecular surfaces do exist. Since the corresponding methods do not require any atom correspondences between molecules, they can be used efficiently for superposing dissimilar molecules. However, this might become a routine technique only if the complicated calculations can be made fast enough to deal with a large number of conformations for each molecule to be superimposed.

References

[1] Kato, Y., Inoue, A., Yamada, M., Tomioka, N., and Itai, A. *J. Comput.-Aided Mol. Design* **6**, 475–486 (1992).

[2] Marshall, G. R., Barry, C. D., Bosshard, H. E., Dammkoehler, R. A., and Dunn, D. A. The conformational parameter in drug design: The active analog approach. In: *Computer-Assisted Drug Design*, ACS Monograph 112. Olsen, E. C., and Christoffersen, R. E. (Eds.). American Chemical Society: Washington D. C. 205–226 (1979).

[3] Dammkoehler, R. A., Karasek, S. F., Shands, E. F. B., and Marshall, G. R. *J. Comput.-Aided Mol. Design* **3**, 3–21 (1989).

[4] Mayer, D., Naylor, C. B., Motoc, I., and Marshall, G. R. *J. Comput.-Aided Mol. Design* **1**, 3–16 (1987).

[5] Kearsley, S. K., and Smith, G. M. *Tetrahedron Comput. Methodol.* **3**, 615–633 (1990).

[6] APEX-3D Package. Biosym Technologies, 10065 Barnes Canyon Road, San Diego, CA 92121, USA.

[7] CATALYST Package. BioCAD Corp., 1390 Shorbird Way, Mountain View, CA 94043, USA.

[8] Martin, Y. C., Bures, M. G., Danaher, E. A., DeLazzer, J., Lico, I., and Pavlik, P. A. *J. Comput.-Aided Mol. Design* **7**, 83–102 (1993).

[9] Kato, Y., Itai, A., and Iitaka, Y. *Tetrahedron* **43**, 5229–5236 (1987).

[10] Manaut, M., Sanz, F., Jose, J., and Milesi, M. *J. Comput.-Aided Mol. Design* **5**, 371–380 (1991).

[11] Clark, M., Cramer III, R. D., Jones, D. M., Patterson, D. E., and Simeroth, P. E. *Tetrahedron Comput. Methodol.* **3**, 47–59 (1990).

[12] Dean, P. M. Molecular recognition: The measurement and search for molecular similarity in ligand–receptor interaction. In: *Concepts and Applications of Molecular Similarity*. Johnson, M. A., and Maggiora, G. M. (Eds.). Wiley: New York; 211–238 (1990).

2.6 The Use of Data Banks

2.6.1 Conversion of 2D Structural Data into 3D Form

An alternative way for generating 3D molecular structures is to start from 2D or 2.5D representations of molecules and to convert this information into a 3D form. While in the sketch approach earlier mentioned a single formula drawing is converted into 3D information, programs like CONCORD [1, 2] and CORINA [3] offer the possibility of employing the structural information for thousands of compounds—which may be stored for example in the databases of pharmaceutical companies—and to convert 2D or 2.5D connection tables into 3D molecular structures. One of these largely commercial databases is available from *Chemical Abstracts* [4].

CONCORD was developed specifically for the 2D to 3D conversion of large database entries containing connection tables of potentially bioactive molecules. For structure generation CONCORD uses a very detailed table of bond lengths. In addition to information such as atomic number, hybridization and bond type, the program regards the "environment" of the atoms included in the bond before assigning bond lengths. This precise selection of bond lengths is especially important for the construction of ring systems, deviations from correct values may have a dramatic effect on the resulting ring conformation.

When starting the 2D conversion the program identifies the so-called "smallest set of small rings". Subsequently, a logical analysis is performed for each particular ring system. Based upon ring adjacency and ring constraints these logical rules decide how the rings will be constructed. In addition, a rough conformation of each ring system is determined, taking into consideration planarity or stereochemical constraints.

If fusion atoms of multicyclic systems are not specified CONCORD creates the isomer with the lowest energy content. After constructing and connecting the ring systems the program modifies the gross conformations in order to remove the internal strain by distributing the strain symmetrically over all atoms in the ring. This procedure leads to cyclic structures with sufficiently relaxed geometries.

The next step in structure generation is to add the acyclic substructures. Bond lengths and bond angles again are taken from predefined tables. To avoid close van der Waals contacts in the built structure the torsion angles are modified in order to obtain energetically acceptable conformations. Besides computational speed the main advantage of CONCORD is that the entire topology of the growing molecule is considered at each step. As a result of this CONCORD yields 3D structures of good quality at low demands of computer time—an important criterion when large databases of 2D information are to be converted into 3D space.

CORINA works in a very similar way to CONCORD. The starting point in creating ring systems is analogous to CONCORD, but CORINA subsequently uses a different approach to connect the ring systems. The rings are fused and the energies of possible ring conformations are calculated using a crude force field. If the actual choice of a particular ring connection was detected as energetically unfavorable, a new attempt is made using other energetically possible conformations of the rings. The generation of ring structures is followed by a geometry optimization step.

Because the energy-based approach of CORINA is less effective than the logical rules used by CONCORD the program is slower in solving the problem of ring connection.

In a similar manner to CONCORD, the acyclic substructures are constructed after the ring system is completed. The chains added to the rings are usually in fully extended conformations. This of course leads to geometries needing further refinement. The torsion angles are rotated until the first conformation is reached, which relieves close contacts. As a result of this rough conformational search the program does indeed provide acceptable structures.

It is important to note that the resulting conformations—only as a matter of chance—either correspond to a conformation in the crystal environment or to a low-energy conformation. The structure finally obtained therefore must be subjected to a conformational analysis in order to detect all possible low-energy conformations.

Both of the programs reviewed are effective alternatives in structure generation, their most significant application being to convert large 2D databases into 3D databases which subsequently are subjected to 3D searching procedures.

2.6.2 3D Searching

3D searching can be of value for several reasons. It may for example be used to produce a set of approximate 3D molecular structures which subsequently can serve as starting points for further investigations like conformational analysis or geometry optimization. If the resulting approximate structures are of sufficient quality they also can be used directly to calculate various properties of the corresponding 3D structures.

It is well known that bioactive ligands which are able to bind to a common receptor must fulfil certain chemical and geometrical criteria. In the case where the 3D coordinates of all atoms of the receptor are known, it is straightforward to search for ligands which mirror the complementary features of the receptor so that they can interact effectively with the binding site. However, in most cases the complete receptor structure is unknown. The lack of precise data on the binding site calls for the definition of a pharmacophore for the drug series under study. For this purpose the limited information on the receptor, together with additional information about some active and inactive ligands to this receptor, is used to generate a crude description of the 3D pharmacophore which can serve as search criteria for 3D searching in large databases.

The 3D search can lead to structures similar to those already known and which satisfy the chemical and geometrical requirements, or can lead to hitherto unknown structures which also possess the features necessary for favorable ligand–receptor interaction.

Prerequisites for effective 3D searching are large databases of 3D structures and suitable software to perform the search. In principle, there are two alternative ways to obtain 3D molecular structures necessary for 3D searching. The easiest of course is to search in existing 3D databases such as the Cambridge Crystallographic Database [5] or the CAS Registry File [6, 7]. These commercial databases offer the possibility either to use already implemented 3D searching programs like GSTAT (which is available from Cambridge Crystallographic Data Centre) or the CAS ONLINE Service from Chemical Abstracts. Other 3D searching programs like 3D search [8] or ALADDIN [9] can also be applied. On the other hand, the user can create his or her own 3D database by converting any 2D database into 3D structural information using programs like CONCORD or CORINA, as already described in section 2.6.1.

A sensitive point in 3D searching is the definition of the 3D search criteria which affects both direction and success of the 3D search. As already mentioned, since exact structural data about receptor proteins in the majority of cases are lacking, the search criteria are usually based on SAR data derived from bioactive compounds.

The "training set" of bioactive structures should contain both active and inactive compounds. By including inactive compounds one can define for example regions which are sterically forbidden for active compounds. The definition of structural requirements for active compounds is straightforward and can be performed on the basis of the 2D molecular structure. For example, if all the active members of the training set contain an acidic substructure or a hydrogen bond acceptor, this information should be used as search criteria. However, it is a more difficult task to define the 3D arrangement of these substructures with respect to each other. When the training set contains fairly rigid active compounds it is of course more simple to define the geometrical requirements. The rigid congeners of the training set can directly serve as templates for the 3D orientation of the substructures, whereas in the case of a set of flexible molecules it is indispensable to determine a common pharmacophore on the basis of a conformational analysis (see also section 2.5).

Usually, three points are used to describe the search query. The pharmacophoric points or substructures are treated as objects and the corresponding interobject distances define the respective 3D orientation. The procedural scenery of 3D searching is nearly identical for most of the programs. In a first step a crude search is used to eliminate compounds which do not meet the 3D search criteria because they lack the required chemical structures in any relative positions. In the second phase the remaining compounds are checked for satisfying both chemical and geometrical demands. As a result of the 3D search a "hit list" is created which contains all the molecules selected in the course of the search.

An effective 3D search should yield a hit list containing a sufficient number of active compounds. The selection of search criteria which are not sufficiently specific will result in an oversized hit list of compounds. Most of the hits will not in fact be able to bind to the receptor because of improper 3D arrangements of the objects. Therefore, the search criteria should be defined in a well-considered manner in order to guarantee a reasonably sized hit list of potentially active compounds.

However, the selection of geometrically very rigid criteria will lead to an unwanted restriction of the search. In order to illustrate this point, let us consider the following situation. The common pharmacophoric pattern derived for all members of the training set does not necessarily correspond to the low-energy conformations of the compounds. Therefore, if the 3D arrangement of the pharmacophore is used as search query, low-energy conformations of active compounds stored in a database may not be selected as hits because they do not fulfil the rigorous geometrical requirements. As a result, active compounds would be lost. In such a situation a more flexible searching procedure would be desirable.

Several attempts have been made to handle this problem. The storage of different low-energy conformations of a compound is unpracticable. As already described in section 2.3.1 the conformational flexibility—and thus the number of different possible conformations—is increasing with the number of rotatable bonds. A modest database already contains hundreds of thousands of compounds. Therefore, the storage of hundreds or thousands of low-energy conformations for each individual compound would consume a huge amount of disk space and the searching time would be absolutely impractical.

One very effective solution to this problem is the introduction of flexible interobject distances [10]. Minimum and the maximum distances which can be achieved by the pharmacophoric points in the molecule are determined. If the molecule is flexible enough to fit all defined, flexible interobject distances simultaneously, then it will be selected and stored.

Nowadays, the 3D searching programs are quite effective and can be used to produce comprehensive hit lists of potentially active compounds. Further refinement of the programs will focus on the improvement in specification of the 3D search criteria to make the search as effective as possible and to implement a more detailed consideration of conformational flexibility. In addition, the development of tools to evaluate the hit lists in a rational way is in progress.

In the framework of this introductory book the increasingly important subject of 3D searching has been presented only very briefly. As this is a relatively new technique significant developments are still in progress. More detailed information can be found in several reviews, such as [2, 11].

References

[1] Pearlman, R. S. *CDA News* **2**, 1–7 (1987).
[2] Pearlman, R. S. 3D Molecular Structures: Generation and Use in 3D Searching. In: *3D QSAR in Drug Design – Theory Methods and Applications.* Kubinyi, H. (Ed.). Escom Science Publishers: Leiden; 41–79 (1993).
[3] Gasteiger, J., Rudolph, C. and Sadowski, *J. Tetrahedron Comput. Methodol.* **3**, 537–547 (1990).
[4] CAS Chem. Registry Syst., Chem. Abstracts Serv., Columbus, OH 43210, USA.
[5] Cambridge Structural Database, Dr. Olga Kennard, F.R.S., Cambridge Crystallographic Data Centre, 12 Union Road, Cambridge CB2 1EZ, UK.
[6] Dittmar, P. G., Farmer, N. A., Fisanick, W., Haines, R. C., and Mockus, J. *J. Chem. Inf. Comput. Sci.,* **23**, 93–102 (1983).
[7] Dittmar, P. G.; Stobaugh, R. E., and Watson, C. E. *J. Chem. Inf. Comput. Sci.* **16**, 111–121 (1976).
[8] Sheridan, R. P., Nilakantan, R., Rusinko, A., Bauman, N., Haraki, K. S., and Venkataraghavan, R. *J. Chem. Inf. Comput. Sci.,* **29**, 255–260 (1989).
[9] Van Drie, J. H., Weininger, D. and Martin, Y. C. *J. Comput.-Aided Mol. Design* **3**, 225–251 (1989).
[10] Guner, O. F., Henry, D. R., and Pearlman, R. S. *J. Chem. Inf. Comput. Sci.* **32**, 101–109 (1992).
[11] Martin, Y. C., Bures, M. G., and Willett, P. Searching Databases of Three-Dimensional Structures. In: *Reviews in Computational Chemistry*, Vol. 1. Lipkowitz, K. B., and Boyd, D. B. (Eds.). VCH: New York; 213–263 (1990).

3 Example for Small Molecule Modeling: Serotonine Receptor Ligands

In this chapter the proceedings leading to the definition of a pharmacophore—based on the construction of a receptor binding site model—will be demonstrated. For this purpose we describe a study on serotoninergic 5-HT$_{2a}$ receptor antagonists which was performed in our laboratory and for which the FITIT method (which has been developed by our group) was used for the superposition procedure. The study employed a simple and straightforward protocol [18] The program fits each energetically accessible conformation of one molecule with each allowed conformation of a second one. The resulting fit pairs are sorted according to rms values and only fit pairs with low rms are saved. This procedure is repeated for the complete list of molecules and in most cases finally yields only a small number of different pharmacophore models. However, this statement can only be true if all known structure–activity relationship data in the series are taken into account and if in addition the conformity of molecular fields for the determined pharmacophoric conformations has been confirmed.

In this example a list of 28 substances with known biological data is used as input. The compounds can be divided into four different structural subsets:

1. 4-(phenylketo)-piperidines
2. tricyclic compounds
3. irindalone compounds
4. butyrophenone derivatives

Typical members of these subsets are shown in Fig. 1.

3.1 Definition of the Serotoninergic Pharmacophore

Unfortunately, the total set does not contain any rigid molecules, but some are at least in part conformationally restricted. This is true for clothiapine and irindalone as well as spiperone, while the members of the ketanserin subfamily which contain five major rotatable bonds show a high degree of conformational freedom.

Experimental structure–activity data for the 5-HT$_{2a}$ antagonist can be summarized as follows. The pharmacological results suggest that two planar aromatic or heterocyclic ring systems in a certain distance connected by an aliphatic or alicyclic chain which contains a basic protonable nitrogen seem to constitute a potent 5-HT$_{2a}$ ligand [1,2]. Additional hydrophobic substituents or a carbonyl group in the heterocyclic ring enhance the antagonistic potency [3].

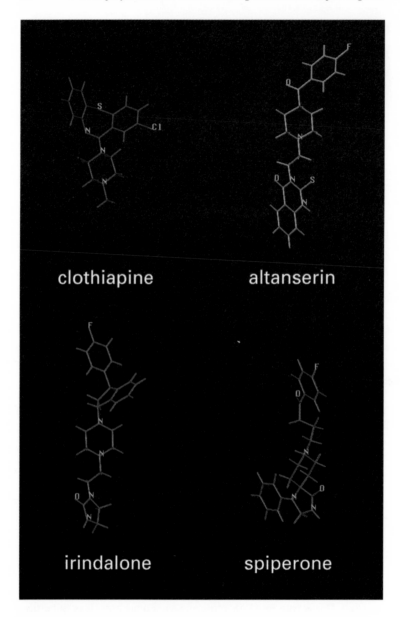

Figure 1. Typical structures of 5-HT$_{2a}$ serotoninergic antagonists.

This knowledge was taken into account in the conformational analysis. In the case considered here an additional advantage could be drawn from the fact that various partial structural elements of the different conformationally constrained molecules can be matched with diverse regions of the highly flexible congeners. In Fig. 2 comparable structural elements of the four

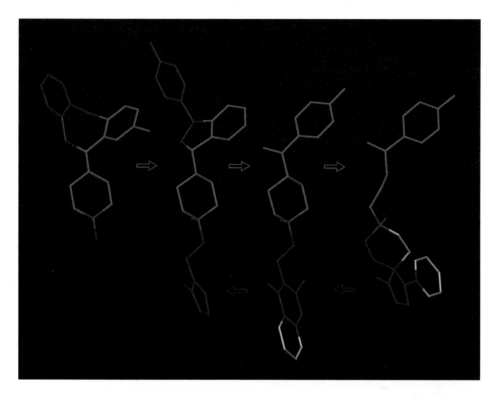

Figure 2. Similarity of the partial structure elements used in the search for a pharmacophoric superposition. Equivalent elements are highlighted by color coding. Molecules from the left: clothiapine, irindalone, altanserin, spiperone.

main groups of 5-HT$_{2a}$ antagonists are color-coded and the stepwise superpositioning procedure followed is indicated.

This type of routine can of course only yield a purely sterical or volume superposition. However, since the interaction of a ligand with its receptor is directed by electronic features it has to be checked whether the discovered pharmacophoric overlap also describes similarity on these grounds. A preliminary approximation can be made comparing the molecular electrostatic potentials (MEPs). In this case it was done on the basis of corresponding AM1-derived charges [4]. Fig. 3 presents the result. The high degree of similarity between the four group representatives is evident. However, a closer inspection of the result of the superpositioning operation reveals that two slightly different pharmacophores were found. Both have to be treated as being equally meaningful because rms values and total agreement of the MEPs within the two sets are similar. In such a situation a general decision for one or the other model can only be made with consideration of the 3D structure of the receptor binding site. If such information is missing no decision is possible. However, a thorough examination of the two pharmacophores for 5-HT$_{2a}$ antagonists brought to light one subtle but significant structural divergence. In one of the two models all the protons at the

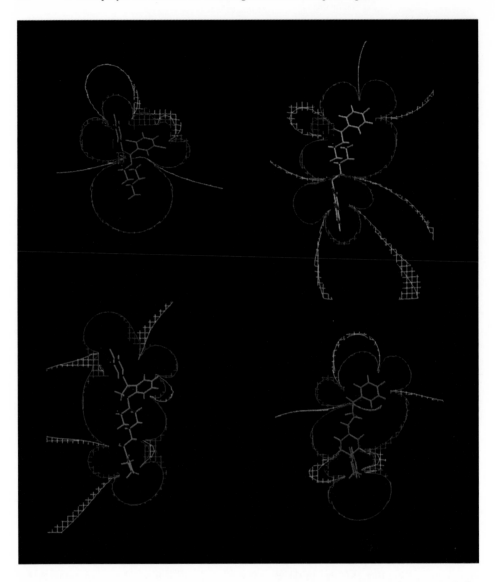

Figure 3. Isopotential contours of altanserin, clothiapine, irindalone and spiperone (counterclockwise from upper right). Color code: blue = –1.0 kcal mol^{-1}, red = 1.0 kcal mol^{-1}, yellow = ± 0.

pharmacophorically important cationic tertiary nitrogens are pointing in the same direction; in the other model this is not the case. Assuming that the cationic protonated nitrogen is involved in a hydrogen-bond-enforced ionic interaction with an anionic receptor binding site, the first pharmacophore model would clearly be favored. Therefore only this pharmacophore will be considered in the further investigation.

3.2 The Molecular Interaction Field

As mentioned and described earlier the evaluation of molecular interaction fields can be performed using GRID [5]. The results of such calculations using two different types of probe are shown in Fig. 4. On one hand a polar hydrogen-bond-active hydroxyl probe was employed and on the other a lipophilic methyl probe has been used. This choice of probes guarantees a rough but rapid determination of basic intermolecular interaction potentials. Using a variety of different probes, a rather detailed picture of the molecular interaction potential for 5-HT$_{2a}$ antagonists can be derived. As shown in Fig. 5 it contains several sites for hydrophobic contacts and hydrogen bond interactions as well as one ionic link.

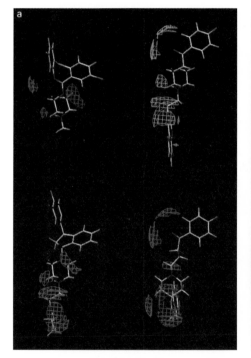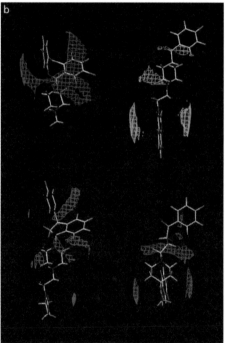

Figure 4. a) GRID contours of the same molecules as in Fig. 3 derived from an aliphatic hydroxyl probe. Energy contoured at –4.0 kcal mol^{-1}. b) GRID contours derived using an aliphatic methyl probe. Energy contoured at –1.4 kcal mol^{-1}.

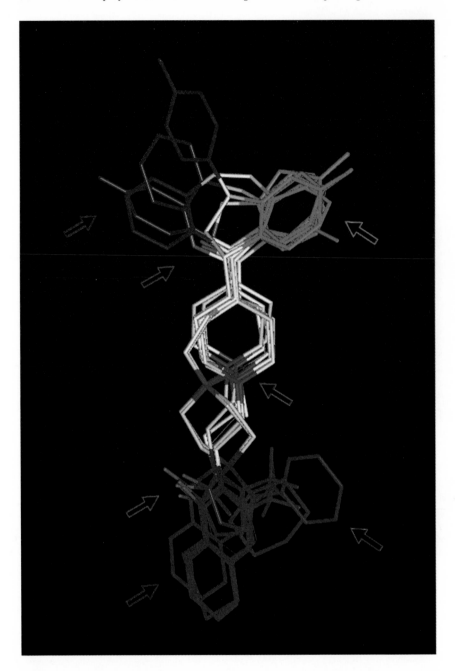

Figure 5. The 5-HT$_{2a}$-antagonistic pharmacophore. Different characteristic regions are marked by colored arrows. Color code: violet = hydrophobic area, green = electron-deficient aromatic system, red = electronegative heteroatoms, pink = protonated nitrogen, blue = large planar ring system (mostly heterocyclic).

3.3 Construction of a 5-HT$_{2a}$ Receptor Binding Site Model

The next step then is to translate this interaction field into a model of the receptor which is composed from single isolated amino acids with chemical properties suitable to satisfy the different types of binding present in the proposed pharmacophore. The relative 3D positions of the amino acid binding sites are defined by the corresponding GRID results. The resulting amino acid receptor model sometimes is called a *pseudoreceptor* [6–8].

A careful inspection of Fig. 5 almost automatically leads to the selection of suitable binding partners needed for the construction of the 5-HT$_{2a}$ receptor map. Hydrophobic amino acids like phenylalanine, tryptophan, valine, leucine or isoleucine should be positioned on both "sides" of the planar cyclic systems. Opposite to the protonated nitrogen an acidic amino acid (e.g. aspartic acid) should be used to fill the location marked by the interaction field created with a hydroxylic probe, whereas the other regions of this field close to the two carbonyl groups found in most of the ligands should be filled with serine, threonine or tyrosine. At this point of course we do not know whether all the discovered interaction possibilities in fact are realized at the receptor level and we will not know this with certainty until the 3D structure of the receptor protein has been elucidated. Nevertheless, structure–activity relationship (SAR) studies are very helpful in order to decide on the existence or absence of binding sites.

In the case under study SAR data tell us that the carbonyl group of the fluorobenzoyl partial structural element is not essential and may be omitted without detrimental effect on the binding strength [9]. Therefore it is concluded that a corresponding hydrogen-bond-donating binding site probably will not be present in the receptor. The same is true for the carbonyl element involved in the heterocyclic system. This can be deduced from the fact that ketanserin derivatives with undiminished affinity are known which present a thiocarbonyl [10] group instead, or even possess a naphthyl system in place of the heterocycle [11]. In conclusion, from the three interaction sites for hydrogen bond contacts between the ligands and the receptor protein, solely the hydrogen-bond-enforced ionic interaction exerted from the protonated nitrogen will be present in the amino acid model.

One additional correction of the interaction field derived receptor map is still necessary. From experimental work it is known that the fluorobenzoyl system may be extensively substituted with hydrophobic elements and that this type of substitution leads to increasing affinity. This fact so far has not been accounted for in the receptor map and therefore we have to add a third hydrophobic amino acid binding site to this region. The final receptor map then appears as in Fig. 6. As mentioned before, the sites A to F of the map now can be occupied by different amino acids with the necessary chemical properties. The available biochemical information, such as amino acid sequence, bacteriorhodopsin homology, alignment studies, etc. has led us to construct the amino acid model of the 5-HT$_{2a}$ receptor presented in Fig. 7.

If experimental knowledge about the amino acid sequence of the receptor protein is absent, it is possible that several different models could be constructed. The decision for one or the other of the hypothetical receptor maps is possible on the basis of calculated interaction energies and their subsequent correlation with the known binding affinities. The model producing the most significant agreement is selected for prediction purposes. Of course, the selection procedure—and coupled to this the quality of the model—is superior if structural information from molecular biochemistry can be used. This for example is true for the

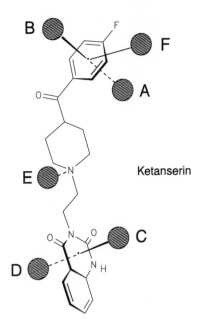

Ketanserin

Figure 6. The receptor map contains six positions for receptor contact. The map has been constructed on the basis of interaction field calculations as well as experimental structure–activity relationship data. Positions A, B, C, D and F depict hydrophobic contacts, position E is an ionic interaction.

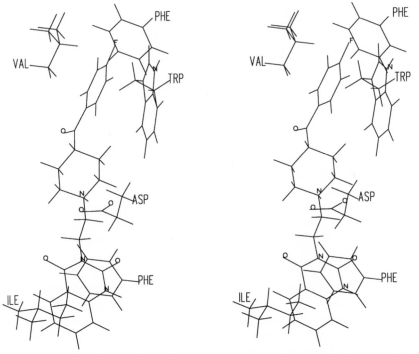

Figure 7. Stereoscopic view of the 5-HT$_2$ receptor model constructed from single isolated amino acids on the basis of the receptor map (see Fig. 6).

G-protein-coupled receptors like the serotoninergic 5-HT$_{2a}$ receptor [12, 13]. Homology searching and sequence alignment operations have been intensively performed, and therefore some ideas about selected amino acids in binding positions in the active site of the receptor do exist [14, 15]. On the basis of this knowledge the derived binding site model very likely reflects important features of the real 5-HT$_{2a}$ receptor protein.

3.4 Calculation of Interaction Energies

Now, the following step is calculation of interaction energies and comparison with experimentally determined binding affinities. This can be done very efficiently using force field methods. For example, the DOCKING procedure and the MAXIMIN module of the SYBYL software package [16] can be employed for optimization of interaction geometries and energy calculation, though other programs can be used equally well. As long as only energy *differences* are of interest the results are quite reliable. However, one must be aware that the force field methods describe only two different types of binding forces adequately, the dispersion and the electrostatic term [17]. The latter depends dramatically on the dielectric constant employed and it is extremely important to choose the correct value in compliance with the respective situation. Inside a protein environment—for example, in the core of the G-protein-coupled receptor channel—the prevailing dielectric constant assumingly is between 3 and 5. Binding sites at protein surfaces are better treated with a value around 10. The constant for in vacuo conditions should be used only in special cases. For example, this would be reasonable if one assumes hydrogen bonds to be of crucial importance for the binding. Since force fields can only simulate the electrostatic part of hydrogen bonds, and neglect the covalent part, this drawback can be partly compensated for by overestimating the electrostatic interaction. Other energy terms which are not included in force field interaction energies are, for example, polarization or charge transfer terms. If there is evidence for an important function of charge transfer processes from SAR data, a corresponding correction of the force field interaction energies should be performed on the basis of quantum chemical calculations.

Interaction energies are determined according to the formula:

$$IE = E_{RL} - (E_R + E_L) \tag{1}$$

where *IE* is interaction energy, E_{RL} is the energy of the receptor–ligand complex, E_R is the energy of the isolated receptor protein, and E_L is the energy of the isolated ligand.

In order to obtain comparable energy data the interaction geometries of the complexes must be generated for all the ligands in an absolutely corresponding manner. All ligands are kept in the pharmacophoric conformation and location. Hydrophobic and polar amino acids mimicking equivalent receptor binding sites are positioned according to the GRID contours. Each individual receptor model–ligand complex is then geometry-minimized in the MAXIMIN force field. No constraints are employed. The procecure therefore simulates an induced fit between ligand and receptor which can be assumed to occur likewise in reality. An energy cut-off of 0.01 kcal mol^{-1} should be used. The calculated energies of interaction significantly must correlate with biological data if the developed receptor binding site model

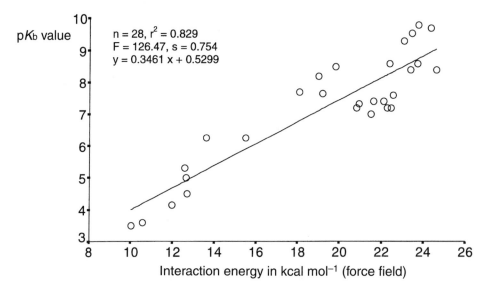

Figure 8. Correlation between calculated interaction energies and experimentally derived binding affinities (pK_b). Interaction energies were obtained on the basis of the receptor binding site model using the SYBYL Maximin force field.

actually does reflect the important features of the real receptor active site. In the case under study this in effect is true. As can be deduced from Fig. 8 the calculated interaction energies correlate significantly with biological data [18]. Some 83% of the variation in the biological data can be explained on the basis of the binding site model; therefore this model is stable enough to be used for the prediction of structurally new 5-HT$_{2a}$ receptor antagonists.

3.5 Validation of the Model

One word of caution is necessary with respect to the experimentally derived biological activities. These by all means should constitute pure receptor binding affinities and ideally should stem from one laboratory. Since the computer models simulate molecular interaction events in a highly simplified manner, the experimental data that are combined with them in a correlation equation must be as close to the molecular niveau as possible. It is therefore absolutely forbidden and indeed virtually nonsense to correlate calculated interaction energies with pharmacological in vivo (whole-animal) data, because the receptor interaction can be blurred or even completely hidden by pharmacokinetics and biotransformation of the drug molecules. Sometimes even functional in vitro data are dangerous if a reaction cascade separates the measured event from the receptor binding interaction.

If receptor map and interaction complex have been generated carefully an approximate but nevertheless correct picture of reality may be obtained. However, as long as the real receptor remains unknown the efficiency and meaning of the model cannot be assessed, except by prediction of new active substances. This should always be the ultimate test of usefulness for each hypothetically derived receptor map.

References

[1] Watanabe, Y., Usui, H., Shibano, T., Tanaka, T., and Kanao, M. *Chem. Pharm. Bull.* **38,** 2726–2732 (1990).

[2] Kennis, L.E.J., Van der AA, M.J.M., Van Heertum, A.H.M.,and Jones, A.J. *Ketanserin Patent Janssen Pharmaceutica N.V., European Patent Office* Nr.0013612, Appl. Nr.803000595 (1980).

[3] Glennon, R.A. Serotonin Receptor Subtypes: Basic and Clinical Aspects. In: *Receptor Biochemistry and Methodology*, Vol. 15. Peroutka, S.J., Venter, J.C.,and Harrison, L.C. (Eds.). Wiley-Liss, Inc.: New York; 19–64 (1991).

[4] AM1 – MOPAC: QCPE-Program No. 455, Version 6.0, Quantum Chemistry Program Exchange, Indiana University, Bloomington, Indiana (USA).

[5] GRID – Molecular Discovery Ltd., Oxford, England

[6] Snyder, J.P., Rao, S.N., Koehler, K.F., and Pellicciari, R. Drug Modeling at Cell Membrane Receptors: The Concept of Pseudoreceptors. In: *Trends in Receptor Research*. Angeli, P., Gulini, U., and Quaglia, W. (Eds.). Elsevier: Amsterdam; 367–403 (1992).

[7] Snyder, J.P., and Rao, S.N. *CDA News* **4/10,** 1/13–15 (1989).

[8] Rao, S.N., and Snyder, J.P. *Cray Channels* **11,** 4–12 (1990).

[9] Herndon, J.L., Ismaiel, A., Ingher, S.P., Teitler, M., and Glennon, R.A. *J. Med. Chem.* **35,** 4903–4910 (1992).

[10] Bogeso, K.P., Arnt, J., Boeck, V., Christensen, A.V., Hyttel, J., and Jensen, K.G. *J. Med. Chem.* **31,** 2247–2256 (1988).

[11] Elz, S. unpublished results.

[12] Guan, X.-M., Peroutka, S.J., and Kobilka, B.K. *Mol. Pharmacol.* **41,** 695–698 (1992).

[13] Kao, H.-T., Adham, N., Olsen, M.A., Weinshank, R.L., Branchek, T.A., and Hartig, P.R. *FEBS Lett.* **307,** 324–328 (1992).

[14] Humblet, C., and Mirzadegan, T. *Annu. Rep. Med. Chem.* **27,** 291–300 (1992).

[15] Trumpp-Kallmeyer, S., Hoflack, J., Bruinvels, A., and Hibert, M. *J. Med. Chem.* **35,** 3448–3462 (1992).

[16] SYBYL, Tripos Associates Inc., St. Louis, Missouri (USA).

[17] Andrews, P.R. Drug–Receptor Interactions. In: *3D QSAR in Drug Design – Theory Methods and Applications*. Kubinyi, H. (Ed.). ESCOM Science Publishers B.V.: Leiden; 13–40 (1993).

[18] Höltje, H.-D., and Jendretzki, U. K. *Arch. Pharm. – Pharm. Med. Chem.* **328,** 577–584 (1995).

4 Introduction to Protein Modeling

4.1 Where and How to get Information on Proteins

Within this book we have, until now, been discussing small molecules. In the second part of the book, the topic of discussion will be *biopolymers*. Since most of the receptors and target molecules known are polypeptides, the main part of the discussion will center on the modeling of proteins.

Each modeling study depends heavily on the quality of the available experimental data, which always serve as the basis of a hypothetical model. Therefore, the first step should always be a careful literature search in order to get a clear picture about the level of knowledge on the biopolymer structure of interest. Valuable information would for example be the complete 3D structure of the receptor or enzyme, ideally derived from crystal data or NMR measurements. After refinement such a structure can be used directly to calculate different properties of the protein or to investigate possible ligand–protein interactions. Unfortunately this situation is still a rare event and in most cases only information on the primary structure of proteins is available.

Besides studying the literature it is very useful to scan different databases to search for primary, secondary and tertiary structural data. Since the number of published sequences and structural information is increasing rapidly an efficient search can only be done by using computer software suitable for this purpose. One such well-known system is the UWGCG program [1] offered by the Genetic Computer Group, Wisconsin. This package allows work with several databases which can be used for the search of an individual protein or DNA structure. The search can be accelerated and specified by employing keywords like author names, journals or families of proteins. Very similar in organization and handling and related to the UWGCG program is the HUSAR program (Heidelberg Unix Sequence Analysis Resources) implemented in the GENIUSnet (Genetic Interactive Unix System). GENIUSnet is a service offered by the German Cancer Research Centre in Heidelberg. In HUSAR a variety of programs greatly facilitates the search in about 20 different sequence and structural databases and the access to up-to-date information on protein sequences.

The collection of available databases in HUSAR comprises for example the EMBL database [2] for nucleic acids, the SWISSPROT [3] and the PIR [4] databases for proteins. Information from the EMBL database (European Molecular Biology Laboratory, Heidelberg) are not only implemented in the GENIUSnet but can also be accessed directly.

SWISSPROT is a comprehensive sequence database which offers a high level of information including a description of the function of a protein, the structure of its domains, etc. The PIR database is related to SWISSPROT. A small part of the information in SWISSPROT is an adaption of the data contained in the protein sequence database of the Protein Information Resource (PIR).

Generally the databases and the appropriate programs are free for use to academic institutions and can be obtained using the file transfer protocol (ftp) from several servers. Some information, however, may only be accessible after a licence agreement has been signed, while some of the programs and information are only available commercially.

The most important and standard database for all structural information on macromolecules is the Brookhaven database [5] which also is available via World Wide Web (http://www.pdb.bnl.gov). In the Brookhaven database atomic coordinates of protein or DNA structures are collected. Because of the continuously growing number of experimentally resolved structures the database is regularly updated. Information hunting in the Brookhaven database can be performed by specifying particular keywords: the author name, a journal, or a part of a sequence for example can serve as a search subject.

Based on the Brookhaven database some smaller structural databases have been created. One is the HSSP database [6] which contains homology-derived structures of proteins. This database combines information from the Brookhaven database and sequences of proteins derived from a sequence database like SWISSPROT.

In general the format, organization and information contained in different structural data files is very similar. As the Brookhaven database is widely used the standard format of a Brookhaven data file will be described in detail in the following (see also Appendix 2). The header of the data file comprises some general information about the protein. It includes the official name, references, resolution of the crystal structure and some useful remarks about the secondary structure composition of the protein. Adjacent to the header are listed the atomic coordinates. Atoms belonging to standard amino acid residues are labelled as ATOM. In order to distinguish between individual peptide chains the ATOMS are separated by an additional line starting with the abbreviation TER. Between ATOMS a bond is generally built when the file is read into the modeling program. This is important as the atoms which do not belong to standard amino acid residues are labelled as HETATM. No connectivity is established between HETATMS. Therefore an additional connectivity table is included at the end of the data file. It is advisable to be careful at this point because it is program-dependent whether or not HETATMS are displayed properly and connectivities are correctly assigned.

HETATMS can either belong to non-standard amino acids or, in the case of complexes, to the ligand molecule involved in the ligand–protein interaction. As the proposed atom type assignment is often incorrect it is absolutely necessary to check carefully all atom types to avoid mistakes resulting in wrong geometries of the ligands (this has already been discussed in Chap. 2.2.1.2).

Usually, all structures from the Brookhaven database do not include hydrogen atoms. For some types of investigation hydrogen atoms can be neglected but for the study of ligand–protein interactions it is inevitable to add the hydrogens. The ligand molecules have to be checked especially carefully in order to confirm that the correct degree of protonation has been assigned in the case of acidic or basic substances.

In addition hydrogen atoms never are allocated to all water molecules. As a consequence they are displayed only as single points representing the oxygen positions. Water molecules can present either simple crystal water distributed near the surface of the protein, or they can be located in the active site. In the latter case it is absolutely necessary to include their complete coordinates into further investigations because they can crucially influence the conformation of the active site structure. This is also true for cations implemented in the crystal structure as they can play an important role for ligand binding or enzyme activity if they are located in the active site.

Most of the modeling programs are able to read the Brookhaven data files without problems and to transform the structural information into a 3D picture of the protein. However, some points of caution should be kept in mind when using experimentally derived information.

In principle, the resolution of a crystal structure should be at least between 2.5 Å and 1.5 Å or better, otherwise the structural information is not very valuable. The purification process of proteins is a difficult and time-consuming task and it may happen that as a result of proteolytic activity some information could be lost before the crystallization process has finished. Therefore amino residues may sometimes be missing, leading to incomplete information contained in the data file.

Some enzymes and proteins fulfil their biochemical function only in the dimeric or trimeric form. The modeler should be aware of this fact because it makes no sense to investigate the functionality of the active site of an enzyme which consists of a dimer when only the monomer structure is present in the Brookhaven file.

Recently the NMR technique has become a frequently used method for obtaining structural information on proteins. NMR has a special bearing on cases where a protein has withstood all efforts to grow sufficiently large crystals. An additional advantage of NMR-derived data is that the conformation of the protein is not influenced by packing forces of the crystal environment. As the NMR measurements are performed in solution the results are highly dependent on the solvent. Experiments in apolar solvents for example lead to an overestimation of hydrogen bonding phenomena. Measurements in aqueous environment should yield a more realistic picture of the protein structure.

The pool of information on proteins is already immense and is growing continuously. However, most of the available databases still only contain information on primary structures. In order to obtain a 3D protein model from these data the application of alignment techniques, knowledge-based and homology modeling approaches is necessary. A detailed discussion on these subjects will be given in section 4.

References

[1] Devereux, J., Haeberli, P., and Smithies, O. *Nucleic Acids Res.* **12**, 387–395 (1984).
[2] Emmert, D. B., Stoehr, P. J., Stoesser, G., and Cameron, G. N. *Nucleic Acids Res.* **22**, 3445–3449 (1994).
[3] Bairoch, A., and Boeckmann, B. *Nucleic Acids Res.* **22**, 3578–3580 (1994).
[4] George, D. G., Barker, W. C., Mewes, H.-W., Pfeiffer, F., and Tsugita, A. *Nucleic Acids Res.* **22**, 3569–3573 (1994).
[5] Bernstein, F. C., Koetzle, T. F., Williams, G. J. B., Meyer, E. F., Brice, M. D., Rodgers, J. R., Kennard, O., Shimanouchi, T., and Tasumi, M. *J. Mol. Biol.* **112**, 535–542 (1977).
[6] Sander, C., and Schneider, R. *Nucleic Acids Res.* **22**, 3597–3599 (1994).

4.2 Terminology and Principles of Protein Structure

The complex 3D structure of proteins can be characterized in four general levels of structural organization: primary, secondary, tertiary and quaternary structure.

1. The primary structure represents the linear arrangement of the individual amino acids in the protein sequence.
2. The secondary structure describes the local architecture of linear segments of the polypeptide chain (i.e. α-helix, β-sheet), without regarding the conformation of the side chains. Another level of structural organization, which was introduced not before very recently, is the so-called *supersecondary structure*. It describes the association of secondary structural elements through side chain interactions. Another term for the same matter is "motif".
3. The tertiary structure portrays the overall topology of the folded polypeptide chain.
4. The quaternary structure describes the arrangement of separate subunits or monomers into the functional protein.

Owing to the remarkable capability of polypeptide chains not only in vivo but also in vitro to fold into functional proteins, it is currently accepted that most aspects of protein architecture and stabilization directly derive from the properties of the particular sequence of amino acids that make up the polypeptide chain (i.e. the primary structure). These properties include the individual characteristics of the side chains of every residue and the influence of the polypeptide backbone on the protein conformation. Only on the basis of this information can 3D structure of a protein be understood. It is not the scope of this introduction to provide a detailed description of all the properties which determine the conformation of a protein, but to explain the main features necessary to understand the contents of the following sections. For a comprehensive description of the principles of protein conformation, the reader is referred to the literature [1–4].

4.2.1 Conformational Properties of Proteins

Generally, only 20 different amino acids are found in naturally occuring proteins. The physico-chemical properties of their side chains, such as size, shape, hydrophobicity, charge and hydrogen bonding, span a considerable range. They avoid, however, the extremes of high chemical reactivity and also, except for proline, strongly restricted degrees of freedom. The question most relevant in view of the 3D shape of proteins is, how the individual side chains interact with the backbone as well as with one another, and what roles they play within particular types of secondary and tertiary structures. The predominant influences of the sequential order on protein conformation are (aside from the linear connectivity and the steric volume) the hydrogen bonding capabilities and the chirality of all (except glycine) amino acid residues. All 19 chiral amino acids possess the L-configuration or according to the Cahn–Ingold–Prelog scheme the S-configuration, with the exception of L-cysteine, which due to a change in ligand priority possesses the R-configuration.

An important convention needed for understanding much of the information available for a particular protein, is the designation of the individual atoms and structural elements of a protein. All atoms, angles and torsion angles that describe the 3D structure of a protein are named using letters in the Greek alphabet. The central carbon atom in amino acid residues is termed α, and the side chain atoms are commonly designated $\beta, \gamma, \delta, \varepsilon$, and ζ in alphabetical order starting from the α carbon atom. The backbone of a protein consists of a repeated sequence of three atoms, belonging to one amino acid residue—the amino N, the C^α and the carbonyl C; these atoms are generally represented as N_i, C_i^α, and C_i' respectively, where i is the number of the residue, starting from the amino end of the chain. As an example, a portion of the backbone of a polypeptide chain is shown in Fig. 1. This illustrates the conventions used in describing protein conformation.

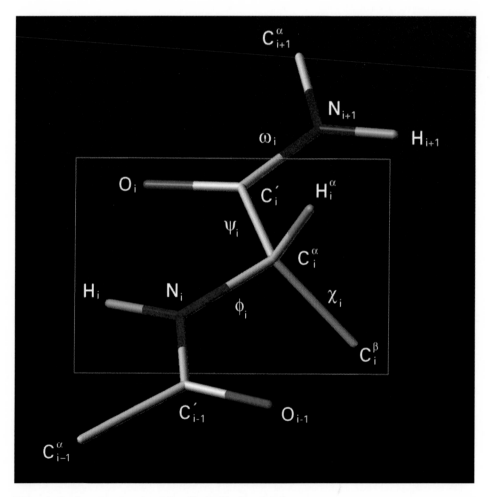

Figure 1. Designation of atoms and torsion angles in a protein.

The main chain torsion angles in proteins are named ϕ (phi), ψ (psi) and ω (omega). Rotation about the N–C^α bond is described by the torsion angle ϕ, rotation about the C^α–C′ bond by ψ, and rotation about the peptide bond C′–N by ω. Torsion angles of the side chains are designated by χ_j (chi$_1$, chi$_2$, etc.) where j is the number of the bond counting outward from the C^α-atom.

The peptide bond is usually planar because of its partial double bond character and nearly always has the *trans* configuration ($\omega = 180°$) which is energetically more favorable than *cis* ($\omega = 0°$). The *cis* configuration sometimes is found to occur with proline residues (about 10%). Small deviations from planarity of the *cis* or *trans* form with $\Delta\omega = -20°$ to $10°$ seem to be energetically acceptable.

The variations of ϕ and ψ are constrained geometrically due to steric clashes between neighboring but non-bonded atoms. The permitted values of ϕ and ψ were first analyzed and determined by Ramachandran et al. [5]. In their work computer models of small peptides were used to vary systematically ϕ and ψ with the purpose of detecting stable conformations. Each conformation, represented by a particular ϕ, ψ combination, was examined for close contacts between atoms. In this rough model the atoms were treated as hard spheres with fixed geometries for the bonds. Only values of ϕ and ψ, for which no close contacts between atoms have been discovered, are permitted and usually are presented in a 2D map, the so-called Ramachandran plot. Since ϕ and ψ constitute a virtually complete description of the backbone conformation, the 2D Ramachandran plot is an important and easy-to-analyze test for the validity of 3D protein structure.

The Ramachandran plot of polyalanine is shown as an example in Fig. 2. The area outside the solid lines corresponds to conformations where atoms in the polypeptide chain are located in distances closer than the sum of their van der Waals radii. These regions are sterically disallowed for all amino acids, except glycine. Glycine, which lacks a side chain, is evenly distributed over the complete plain of the Ramachandran plot. The shaded regions correspond

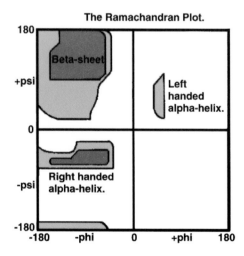

Figure 2. Ramachandran plot of a polyalanine.

to conformations where no steric clashes are found, i.e. these are the allowed regions (or favored regions). The area directly outside the boundaries of this region includes conformations which are permitted if slight alterations of bond angles are accepted. Ramachandran plots for other amino acids appear comparable in the shape of the various regions.

Sub-regions of ϕ,ψ space are generally named after the secondary structure elements which result when the corresponding ϕ,ψ-angles occur repeatedly. The right-handed α-helix for example resides in the lower left near $-60°, -40°$, the broad region of extended β-sheets in the upper left around $-120°, 140°$, and the slightly unfavored left-handed α-helical region in the upper right near $60°, 40°$. Conformational properties and other relevant parameters of these secondary structures are described in the following sections.

4.2.2 Types of Secondary Structural Elements

4.2.2.1 The α-Helix

The right-handed α-helix is the best known and most easily recognized secondary structural element in proteins [6, 7]. Approximately 32% to 38% of the residues in known globular proteins are involved in α-helices [8]. α-Helices are classified as repetitive secondary structure. That is, all C_α-atoms of α-helical amino acids are in identical relative positions. Thus, the ϕ,ψ torsion angle pairs are the same for each residue in the helix. The structure of an α-helix repeats itself every 5.4 Å along the helix axis; this means that the α-helices have a pitch of p = 5.4 Å. α-Helices have 3.6 amino acid residues per turn, i.e. a helix of 36 amino acids would form 10 turns.

The α-helical structure is mainly formed and stabilized by repeated hydrogen bonds between the carbonyl function of residue n and the NH of residue $n+4$ (see Fig. 3). This results in a very regular and energetically favored state. α-Helices observed in protein structures are always right-handed. L-amino acids cannot form extended regions of left-handed α-helix because the C^β-atoms would collide with the following turn. Only individual residues are found which possess the ϕ,ψ torsion angles of a left-handed α-helix. So, when speaking of an α-helix, usually the right-handed α-helix is meant.

The exact geometry of the α-helix is found to vary somewhat in natural proteins, depending on its environment. The ideal α-helix ($\phi = -57°$ and $\psi = -47°$) is only one version of a family of similar structures [6]. More usually, a slightly different α-helix geometry ($\phi = -62°$ and $\psi = -41°$) can be observed in proteins [7]. This conformation is more favorable than the ideal α-helix because it permits each carbonyl oxygen to make hydrogen bonds to both the NH of residue $n+4$ and the aqueous solvent (or other hydrogen bond donors).

The side chains of an α-helix are pointing outwards into the surrounding space. Several restrictions exist for side chain conformations, especially for side chains with branched C^β-atoms (Val, Ile, Thr). Proline residues normally are incompatible with the α-helical structure because, due to the cyclic structure, the amide nitrogen lacks the hydrogen substituent necessary for hydrogen bonding. If single proline residues nevertheless appear in long α-helices (e.g. in some of the transmembrane α-helices of bacteriorhodopsin), this appearance yields a local distortion of the α-helical geometry.

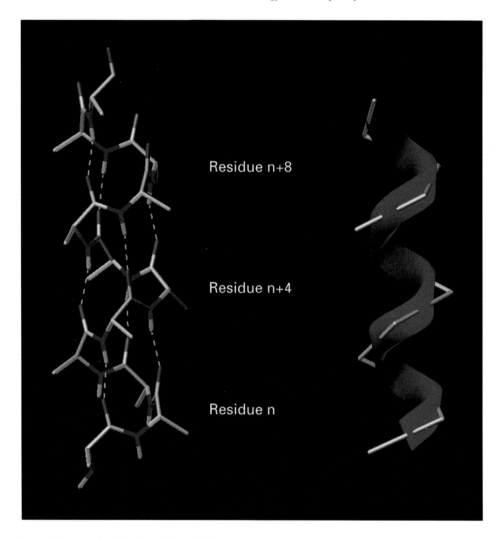

Figure 3. General architecture of an α-helix.

Variations of the classical α-helix in which the protein backbone is either more tightly or more loosely coiled (with hydrogen bonds to residues $n+3$ and $n+5$), are named 3_{10}-helix and π-helix, respectively. In general, these helix types play only a minor role in the architecture of proteins. However, 3_{10}-helices frequently form the last turn of a classical α-helix.

4.2.2.2 The β-Sheet

Besides the α-helix, the second most regular and recognizable secondary structural motif is the β-sheet [9, 10]. Like the α-helix it is a periodic element. β-Sheets are formed from β-strands which develop when a linear extended conformation of a polypeptide backbone ($\phi = -120°$, $\psi = 140°$) appears [9]. Since interactions between residues of the same strand are not possible, a β-strand is only stable as part of a more complex system, the β-sheet. As in α-helices all hydrogen bond donor or acceptor groups of the peptide backbone are engaged in the formation of hydrogen bonds, however, because these bonds appear not intra- but intermolecular, β-strands are energetically less favored. In contrast to α-helices—which consist of a singular stretch of directly bound residues—β-sheets possess a much more pronounced structure modulating effect because they are composed of several β-strands which can be distributed over a large part of the sequence.

Adjacent β-strands can be arranged in either parallel or antiparallel fashion. In parallel sheets the strands all run in the same direction (see Fig. 4(a)), whereas in antiparallel sheets they run in opposite directions (see Fig. 4(b)).

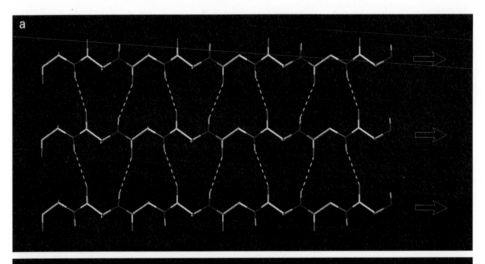

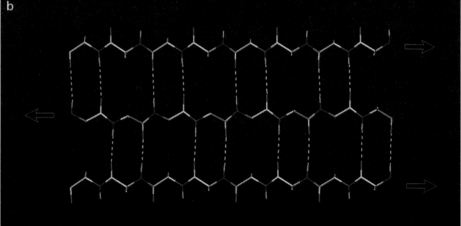

Figure 4. Architecture of a) parallel and b) antiparallel β-sheets.

The side chains of β-strands are located nearly perpendicular to the plane of the hydrogen bonds (between the single strands). Along the strand they alternate from one side to the other. For antiparallel β-sheets typically one side is buried in the interior of the protein and the other side is exposed to the solvent. Therefore, the physico-chemical character of the amino acids tends to alternate from hydrophobic to hydrophilic. Parallel β-sheets, on the other hand, are usually buried on both sides, so that the central residues tend to be hydrophobic, and hydrophilic amino acids are found abundantly towards the ends. For both types of β-sheet, edge strands can be much more hydrophilic than central strands.

β-Sheets are very common in globular proteins (20–28%) [8]. They can consist exclusively of parallel or antiparallel strands or are formed from a mixture of both. Purely parallel sheets are less frequent, while purely antiparallel sheets are very common. Antiparallel sheets often consist of as few as two or three strands, whereas parallel sheets always have at least four. Mixed sheets usually contain 3–15 strands.

4.2.2.3 Turns

Approximately one-third of all residues of globular proteins are involved in turn regions. The general function of turns is to reverse the direction of the polypeptide chain. Often turns are located on the protein surface and therefore contain predominantly charged and polar amino acids.

Various different types of reverse turns have been observed in proteins. Their specific features depend for example on the type of secondary structural motifs which are linked by them. For a detailed description of all observed turn types the reader is referred to the literature [1–4, 11, 12].

Turns often connect antiparallel β-strands. In this case they are named β-turns or hairpin bends [12]. Some 70% of hairpin turns are shorter than seven residues in length; most often they include only two residues. Larger loops have less well-defined conformations, which are often influenced by interactions with the rest of the protein. In all reverse turns the peptide groups are not paired by regular hydrogen bonds, but are accessible to the solvent. For this reason reverse turns often appear on the protein surface.

In general, the periodic secondary structural elements in proteins (α-helices and β-sheets) are rather short. The length of an α-helix is usually 10–15 residues (12–22 Å). A single β-strand is found to count 3–10 residues (7–30 Å). Most of the described ideal geometries of helices and sheets are only rarely observed in nature. Often, the geometries of secondary structures are more or less distorted. For example, solvent-exposed α-helices very often show a curved helix axis. Most β-sheets in folded proteins are rather twisted than planar with a twist of 0–30° between the single strands.

The common properties of proteins described here provide only some general rules of protein architecture. Each naturally occurring protein, on the other hand, is unique and attains its functional and structural character by means of specific non-covalent interactions. It is therefore necessary to compare each computer-generated structure with "real" 3D structures of proteins, and to include as much as possible information about protein structures in the process of protein modeling.

The exclusive presentation of secondary structural motifs of a complex protein in a schematic form is a very helpful tool for comprehending the overall structure. Usually in this kind of representation the side chains are omitted to yield a clearer picture of the whole protein, including the various secondary structural elements. Helices are often described by cylinders or coiled ribbons and extended strands of β-sheets by broad arrows indicating the amino-to-carboxy direction of the backbone. The 3D structure of triose phosphate isomerase is presented in such simplified form in Fig. 5.

4.2.3 Homologous Proteins

It has long been recognized that the evolutionary mechanism of gene duplication which is associated with mutations, leads to divergence and thereby to the foundation of families of related proteins with similar amino acid sequences and similar 3D structures. The proteins that have evolved evolutionarily from a common ancestor are said to be *homologous*. Two homologous sequences can be nearly identical, similar to varying degrees, or dissimilar because of extensive mutations. As a matter of fact the sequence similarity in homologous proteins is less preserved than the structural similarity. Or stated in a different way, 3D structures of homologous proteins have been remarkably conserved during their evolution because the common structure is crucial for the specific function of the proteins. The

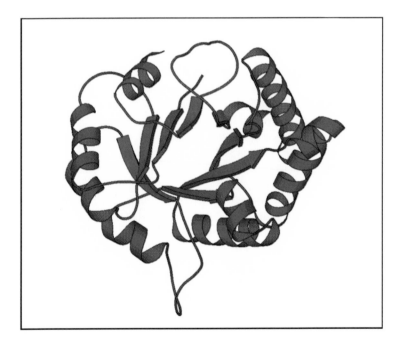

Figure 5. The 3D structure of triose phosphate isomerase presented in simplified form using MOLSCRIPT [13].

conservation of protein structure has been detected in many protein families. The 3D structures of α-chymotrypsin and trypsin, belonging to the family of trypsin-like serine proteases, can be cited as an example. They are remarkably similar, although they share only 44% identical amino acid residues. This topological similarity can easily be observed in Fig. 6. Other members of the family of serine proteases have changed more drastically during evolution. Bacterial serine proteases for example show only 20% sequence identity when compared with the mammalian enzymes like thrombin, trypsin or chymotrypsin. However, if the 3D structural similarity is considered the main features are still present.

The question which immediately comes to mind in this respect is how such large dissimilarities in the primary sequences are compatible with the observed structural similarity. The answer was found empirically and can be summarized as follows. The most pronounced dissimilarities generally appear in regions close to the protein surface, the so-called loop regions. In these regions even the physico-chemical properties of the side chains have often changed. Residues located in the interior of proteins, however, vary less frequently and less distinctly. This leads to the situation that generally a common core of residues comprising the center of the protein and the main elements of secondary structure remain highly conserved within a family of homologous proteins.

Within homologous proteins the elements of secondary structure can move relative to each other, can change in length, or can even disappear completely. However, an α-helix is not usually replaced by a β-sheet, or vice versa. In general, neither the order nor the orientation (parallel or antiparallel) of β-strands has ever been recognized to differ between proteins of the same family.

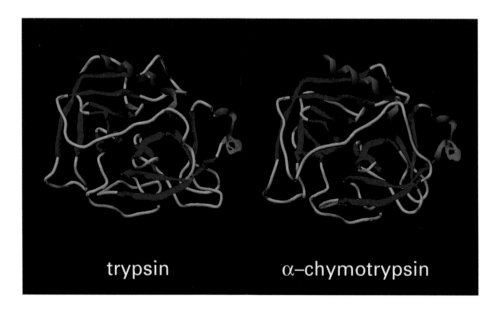

Figure 6. 3D structure of two homologous enzymes. Color code: red = α-helix; blue = β-strand; yellow = peptide backbone.

In summary, the overall conformations of homologous proteins appear to have been highly conserved during evolution. This fact forms the basis for the development of the knowledge-based approach in protein modeling, which will be described in the next section.

References

[1] Creighton, T. E. *Proteins: Structures and Molecular Properties,* 2nd Ed. W. H. Freeman and Company: New York 1993.

[2] Branden, C., and Tooze, J. *Introduction to Protein Structure.* Garland Publishing Inc.: New York 1991.

[3] Schulz, G. E., and Schirmer, R. H. *Principles of Protein Structure.* Springer Verlag: New York 1979.

[4] Fasman, G. D. *Prediction of Protein Structure and the Principles of Protein Conformation.* Plenum Press: New York 1989.

[5] Ramachandran, G. N., and Sasisekharan, V. *Adv. Prot. Chem.* **23**, 283–437 (1968).

[6] Pauling, L., Corey, R. B., and Branson, H. R. *Proc. Natl. Acad. Sci. U.S.A.* **37**, 205–211 (1951).

[7] Barlow, D. J., and Thornton, J. M. *J. Mol. Biol.* **201**, 601–619 (1988).

[8] Kabsch, W., and Sander, C. *Biopolymers* **22**, 2577–2637 (1983).

[9] Chou, K. C., Pottle, M., Nemethy, G., Veda, Y., and Scheraga, H. A. *J. Mol. Biol.* **162**, 89–112 (1981).

[10] Pauling, L., and Corey, R. B. *Proc. Natl. Acad. Sci. U.S.A.* **37**, 729–740 (1951).

[11] Rose, G. D., Gierasch, L. M., and Smith, J. A. *Adv. Prot. Chem* **37**, 1–109 (1985).

[12] Sibanda, B. L., and Thornton, J. M. *Nature* **316**, 170–174 (1985).

[13] Kraulis, P. J. *J. Appl. Crystallogr.* **24**, 946–950 (1991).

4.3 Knowledge-Based Protein Modeling

As we have already discussed in section 4.1 extensive information on primary and secondary structure of proteins are stored in various databases. Protein sequence determination is now routine work in molecular biology laboratories. As a result, the rate of publication of primary sequences has increased dramatically in the last few years. Sequences of more than 100 000 proteins are now available. The translation of sequences into 3D structure on the basis of X-ray crystallography or NMR investigations, however, takes much more time. Therefore, hitherto (by the end of 1995) the 3D structures of not more than 3000 proteins are available in the Brookhaven protein database. In certain circumstances it can take, depending on the kind of proteins studied, more than one year to perform a complete structure determination [1]. Therefore, many more protein sequences are known than complete 3D structures. Because of the technical problems related to experimental 3D structure elucidation, theoretical procedures for predicting protein 3D structure on the basis of the respective amino acid sequence are urgently needed. Since a general rule for the folding of a protein has not yet been developed, it is necessary to base structural predictions on the conformations of available homologous reference proteins [2–4] (see also section 4.2 for the underlying principles).

If one sequence is found homologous to another, for which the 3D structure is available, the knowledge-based approach (also called the homology modeling approach) is the method of choice for predicting the structure of the unknown protein. The underlying idea of homology modeling is to make use of the collected body of knowledge about already resolved proteins. In a first step the sequence of a new protein is compared with all sequences of structurally known proteins stored in a database. Proteins in the database which are identified as homologous to the unknown are retrieved and used as templates for the structural prediction of the unknown protein. This approach was developed by several authors and is described in detail in the following paragraphs [5–8].

Successful knowledge-based model building, however, depends strongly upon how closely the structure that one is attempting to model fits the chosen template [9]. Because, at present, our understanding of protein folding patterns is still rather limited, the only criterion that can be applied for structure prediction is the examination of the extent of sequence homology between known and unknown protein. Although the conclusion of many studies in the past was that structural homology persists even if sequence homology is hardly detectable, for the purpose of knowledge-based modeling the reverse is important. The prediction of structural similarity between different proteins can only be based on the detection of homologies in their sequences. Thus, the comparison of sequences using alignment methods is a central technique in homology modeling and will be described in detail in section 4.3.1.

The process of homology modeling involves the following steps:

1. Determination of proteins which are related to the protein to be studied.
2. Identification of structurally conserved regions (SCRs) and structurally variable regions (SVRs).
3. Alignment of the sequence of the unknown protein with those of the reference protein(s) within the SCRs.

4. Construction of SCRs of the target protein using coordinates from the template structure(s).
5. Construction of SVRs.
6. Side chain modeling.
7. Structural refinement using energy minimization methods and molecular dynamics.

4.3.1 Procedures for Sequence Alignments

The first step in homology modeling is the assignment of the unknown protein structure to a protein family. In many cases this information is already known because the sequence to be modeled belongs to a well-known protein family. However, this may not be true. Then it is necessary to compare the new sequence with thousands of sequences already stored in protein databases and to identify, if possible, homologous ones.

In the past, identifying new proteins through database searches has been difficult and time-consuming. Computer programs required several hours or made far-reaching compromises in sensitivity or selectivity of the search. However, more efficient and rapid searching procedures have been developed in recent years [10–13]. Examples are the FASTP [12] or the BLAST [13] program. Many commercially available software packages (like HOMOLOGY [14], MODELLER [15], COMPOSER [16], WHAT IF [17], UWGCG [18]) include these or comparable programs for automatical database searching.

The central technique used for amino acid sequence comparison is the so-called *sequence alignment*. In the framework of homology modeling the sequence alignment procedure is of importance for several reasons. Firstly it is used to search databases in order to find related sequences and to identify which regions of the detected proteins are conserved, thus suggesting where the unknown protein may also be structurally conserved. This for example can be performed employing the above-mentioned FASTP or BLAST programs. Secondly, sequence alignment is used for detection of correspondences between amino acids of the structurally known reference protein and those of the protein to be modeled. These correspondences are the basis for transferring the coordinates of the reference protein(s) to the model protein. For this task the more sensitive and selective alignment procedures described below are needed.

A very natural procedure for aligning sequences would be to simply write them in tabular form for visual inspection. Of course this would be not only unsystematic, it would be very time-consuming, especially if more than two sequences are to be compared. For that reason many programs have been developed which are able to perform alignments automatically [18–21]. Because the alignment of amino acid sequences is such a crucial step in homology modeling of proteins, many different methods and programs have been published and still are being developed. It is beyond the scope of this book to discuss all of them, but the reader is referred to the literature [12, 13, 18, 19].

One of the earliest attempts to clarify whether the structural similarity existing between proteins is due to homology or occurs by chance, was carried out by Needleman and Wunsch [20]. Variants of the algorithm used by these authors have been further developed independently by others and applied in many fields. These programs are more sensitive in detecting homology than the database search programs, but on the other hand are slower in finding an optimal alignment. However, the great advantage of the Needleman and Wunsch algorithm is that final detection of the best alignment for two sequences is guaranteed. As a consequence computer programs based on this method (ALIGN, BESTFIT and GAP which are included in the UWGCG program package [18]) have been widely used for biological sequence comparison. Whereas the original Needleman and Wunsch algorithm is only able to align two sequences, many up-to-date programs handle the alignment of more than two sequences. These so-called multiple alignment methods are significantly more difficult than the pairwise alignment techniques. This is because the number of possible alignments increases exponentially with the number of the sequences to be compared. Several programs have been derived to provide approximate solution of this problem (for example CLUSTALW [18] or MAXHOM [21])

In contrast to the above-described procedures—which search for the global optimal similarity of sequences—other approaches seek to identify the best local similarities between two sequences. These so-called *optimal local alignment* methods are likewise based on a modified Needleman and Wunsch algorithm and represent an important tool for comparing sequences. This is especially true for the location of highly homologous regions dispersed over long sequences [22–24]. The basic idea of these methods is to consider only relatively conserved subsequences of homologous proteins; dissimilar regions do not contribute to the measure (Fig. 1).

In the course of comparising of two sequences the alignment procedures, at least in effect, seek to duplicate the evolutionary process involved in converting one sequence into another. For this operation a kind of scoring scheme is required that dictates the weight for aligning a particular type of amino acid with another. This type of scoring scheme is provided by the so-called *homology matrices*, which make use of the most probable amino acid substitutions according to physical, chemical or statistical properties. High numerical values in the matrix imply that a substitution is probable, whereas low values indicate that a substitution is unlikely to occur. From the various kind of matrices which are in use [25–29] the most often applied are:

1. *Identity matrix*: this is the simplest matrix that gives a score of 1 to identical pairs and 0 to all others.
2. *Codon substitution matrix*: the scoring values for this matrix are derived from the DNA base triplets coding for the amino acid pairs. For each pair, all of the possible nucleotide triplets are examined and the number of point mutations required to change one amino acid into the other are evaluated. Identical amino acids get a score of 9, one required mutation gives a score of 3, and two mutations yield a score of 1.
3. *Mutation matrix* (also known as the Dayhoff or PAM250 matrix [25]): this matrix was obtained by counting the number of substitutions from one particular amino acid by others observed in related proteins across different species. Large scores are given to identities and substitutions which are found frequently, and low scores are assigned to mutations that are not observed. Due to this procedure larger scores are used for certain non-identical pairs than for some identical ones. The Dayhoff matrix (Fig. 2), is the most widely used scoring scheme. It is often applied for finding an initial alignment for two unknown sequences. An advanced form of the Dayhoff matrix was suggested by Gribskov et al. [26]. The Gribskov matrix assigns the highest score always to identical amino acid pairs.

```
ENTCL :  PVSEKQLAEVVANTITPLMKAQSVPGMAVAIY-QGKPHYYTFGKADIAANKPVTPQTLFELGSISKTFTGVLGGDAIA-
CITFR :  AKTEQQIADIVNRTITPLMQEPAIPGMAVAIY--EGKPYYFTWGKADIANNHPVTQQTLFELGSVSKTFNGVLGGDRIA-
MEN1  :  QTADVQQKLAELERQSG-GRLGVALINTADNSQILYR-----------------ADRFAMCSTSKVMAVAAVLKKSE-
STAU  :  KELNDLEKKYN-AHIGVYALDTKSGKEVKFN-------------------SDKRFAYASTSKAINSAILLEQVP-
BALI  :  DDFAKLEQFD-AKLGIFALDTGTNRTVAYR--------------------PDERFAFASTIKALTVGVLLQQKS-


ENTCL :  -RGEISLDDAVTRYWPQLTGKQWQ------------GIRMLDLATYTAGGLPLQVPDEVTDNASLLRFYQNWQPQWKPGTTRLYANASIGLFGALAVKPSGMPYE
CITFR :  -RGEIKLSDPVTKYWPELTGKQWR-----------GISLLHLATYTAGGLPLQIPGDVTDKAELLRFYQNWQPQWTPGAKRLYANSSIGLFGALAVKSSGMSYE
MEN1  :  -SEPNLLNQRVEIKKSDLVNYNPIAEKHVDGTMSLAELSAAALQ--------------YSDNVAMNKLISHVGGP--ASVT
STAU  :  --YNKLNKKVHINKDDIVAYSPILEKYVGKDITLKALIEASMT------------YSDNTANNKIIKEIGGI--KKVK
BALI  :  ---IEDLNQRITYTRDDLVNYNPITEKHVDTGMTLKELADASLR-------------YSDNAAQNLILKQIGGP--ESLK


ENTCL :  QAMTTRVLKPLKLDHTWINVPKAEEAHYAWGYRDGKAVRVSPGMLDAQAYGVKTNVQDMANWVMANMAPENVADASLKQGIALAQSRYWRIGSMYQGLGW
CITFR :  EAMTRRVLQPLKLAHTWITVPQSEQKNYAWGYLEGKPVHVSPGQLDAEAYGVKSSVIDMARWVQANMDASHVQEKTLQQGIELAQSRYWRIGDMYQGLGW
MEN1  :  AFARQLG-----DETFRLDRTEPTLNTAIPGDPRD------------TTSPRAMAQTLRNLTLGKALG---DSQRAQLVTWMKGNTGAASIQA
STAU  :  QRLKELG-----DKVTNPVRYEIELNYYSPKSKKD-------------TSTPAAFGKTLNKLIANGKLS---KENKKFLIDLMLNNKSGDTLIKD
BALI  :  KELRKIG-----DEVTNPERFEPELNEVNPGETQD--------------TSTARALVTSLRAFALEDKLP---SEKRELLIDWMKRNTGDALIRA


ENTCL :  EMLNWPVEANTVVEGSDSKVALAPLPVAEVNPPAPPVKASWHKTGSTG--GFGSYVAFIPEK----QIGVMLANTSY----PNPARVEAAYHILEAL
CITFR :  EMLNWPLKADSIINGSDSKVALAALPAVEVNPPAPAVKASWHKTGSTG--GFGSYVAFVPEK----NLGIVMLANKSY---PNPARVEAAWRILEKL
MEN1  :  GLPAS------------------------WVVGDKTGSGD-YGTTNDIAVIWPKD-RAPLLVTYFTQPQPKAESRRDVLASAAKIVTNGL
STAU  :  GVPKD------------------------YKVADKSGQAITYASRNDVAFVYPKGQSEPIVLVIFTNKDNKSDKPNDKLISETAKSVMKEF
BALI  :  GVPDG------------------------WEVADKTGAAS-YGTRNDIAIIWPPK-GDPVVLAVLSSRDKDAKYDDKLIAEATKVVMKAL
```

Figure 1. Multiple sequence alignment of cephalosporinases from *Enterobacter cloacae* (ENTCL) and *Citrobacter freundii* (CITFR), with penicillinases from *Escherichia coli* (MEN1), *Bacillus licheniformis* (BALI) and *Staphylococcus aureus* (STAU). Red letters indicate the determined SRCs.

	Ala	Arg	Asn	Asp	Cys	Gln	Glu	Gly	His	Ile	.
Ala	2	-2	0	0	-2	0	0	1	-1	-1	
Arg	-2	6	0	-1	-4	1	-1	-3	2	-2	
Asn	0	0	2	2	-4	1	1	0	2	-2	
Asp	0	-1	2	4	-5	2	3	1	1	-2	
Cys	-2	-4	-4	-5	12	-5	-5	-3	-3	-2	
Gln	0	1	1	2	-5	4	2	-1	3	-2	
Glu	0	-1	1	3	-5	2	4	0	1	-2	
Gly	1	-3	0	1	-3	-1	0	5	-2	-3	
His	-1	2	2	1	-3	3	1	-2	6	-2	
Ile	-1	-2	-2	-2	-2	-2	-2	-3	-2	5	
.											

Figure 2. Dayhoff evolutionary mutation matrix.

4. *Physical property matrices*: the scores of corresponding matrices are based on similarity indices for certain physical properties of amino acids, such as hydrophobicity, polarizability or helical tendency [28].

Differences in sequence lengths or variations in the locations of conserved regions complicate the alignment procedure. If one or both of the mentioned problems are found, gaps are introduced into the sequence to allow the simultaneous alignment of all conserved regions. To limit the total number of inserted gaps (a large number would render the alignment increasingly unrealistic), an additional factor is implemented into the alignment algorithms, the so-named *gap penalty function*. The overall balance between the number of aligned amino acids and the smallest number of required gaps leads to an optimal alignment.

The combination of an alignment algorithm, a scoring matrix, and a gap-weighting function constitutes a system which can optimally align two or more sequences. The quality of a particular alignment is described by the *alignment score*. It is important to know that a derived alignment for related sequences is optimal only for the chosen parameters; changing the values can lead to a different alignment and a different score. Thus, it should be borne in mind that automatic sequence alignment methods are far from being perfect. The resulting alignment should always be verified for reasonableness. All known information on all levels on protein organization (primary, secondary and tertiary structure) have to be incorporated in the examination. Only when the derived alignment agrees with all known structural data can it be used as a basis for the generation of a protein model.

Another fundamental problem of all sequence alignments is found in the fact that recognizable sequence homology is lost more rapidly during evolution than the underlying structural similarity. Thus, it is difficult to give simple rules for the degree of similarity necessary to demonstrate unambiguously that two protein sequences are homologous. This depends strongly on the lengths of the sequences and their amino acid compositions. During the past decade several investigations have been performed to quantify the relation between sequence and structural homology [30–32].

Doolittle has defined some rules of thumb which can ease the decision [30]. If the sequences are longer than 100 residues and are found to be more than 25% identical (with appropriate gaps) then they are very likely related. If the identity is in the range of 15–25%, then the sequences may still be related. If the sequences are less than 15% identical, they are probably not related.

In order to be able to take a decision in the undecided range between 15–25% homology it must be proven that the alignment is statistically meaningful. One way to evaluate this point is by comparing the actual alignment score, which reflects the amount of homology between two sequences, with the average alignment score of randomly permuted sequences (which were generated by randomly exchanging the amino acid residues in the original sequences). This procedure preserves the exact length and amino acid composition of the proteins, and the statistical variation of the random comparison provides a measure of the significance of the observed similarity. A number of n randomizations for both sequence 1 and 2 will be generated. Each derivative of sequence 1 is then aligned against each derivative of sequence 2, resulting in a total of n^2 alignments. Both the mean and the standard deviations of the alignments are normally reported and can be compared with the original score. As an approximate guide; if the alignment score is more than six times the standard deviations above that for the random alignment, most of the residues in secondary structures will be correctly aligned [31].

Chothia and Lesk have performed an investigation on homologous proteins in order to quantify the relation between sequence homology and 3D similarity in core regions of entirely globular proteins [32]. They have found, that the success to be expected in modeling the structure of a protein from its sequence (using the 3D structure of a homologous protein as template) depends to a high degree upon the extent of sequence identity. They concluded that a protein structure provides a close general model for other proteins if the sequence identity is above 50%. If the sequence homology drops to 20%, large structural differences can occur (see Fig. 4 in section 4.5.3). However, they found that the active site of distantly related proteins can have very similar geometries. Thus, in cases where the sequence identities are low, the structure of the active site in a protein may provide a reliable model for those in related proteins.

4.3.2 Determination and Generation of Structurally Conserved Regions (SCRs)

Building a protein model using the homology approach is based on the fact that there are regions in all proteins belonging to the same protein family that are nearly identical in their 3D structures. These regions tend to be located at the inner cores of proteins where differences in peptide chain topology would have significant effects on the tertiary structure of the protein

[33]. Accordingly, it has been observed that the secondary structural units of strongly related proteins, above all α-helices and β-strands, occupy the same relative orientations throughout the whole protein family. As a natural follow-up these regions lend themselves to being used as the basic framework for the assignment of atomic coordinates for one of the other proteins belonging to the same family. These segments are called *structurally conserved regions* (SCRs).

The accurate assignment of SCRs within a family of homologous proteins is affected by several factors. The way to proceed depends on the number of available crystal structures of homologous proteins. It is fortunate when more than one crystal structure at atomic resolution is available. In this situation one can examine all structures in order to discover where the proteins are conserved structurally, even with regard to the 3D structure. To recognize the conserved parts of the proteins, they must be superimposed relative to each other. This is normally done using least-squares fitting methods. The main problem in this context is the selection of the corresponding fitting atoms; this means that it is not known a priori which part of the protein should be aligned to receive the best 3D overlap. In a first approximation the structures can be superimposed by least-square fitting of the C^{α}-atoms [3]. The initial superposition then can be optimized using only matching points located in secondary structural elements that are found to be conserved. Several approaches have been developed which try to solve the fitting problem automatically [34–40].

Matthews and Rossmann [40] have suggested a method which uses the least-squares fitting procedure. In a first step, two protein structures—which have to be aligned—are least-squares fitted using an initial set of equivalent residues. The equivalences are then updated according to both the distances between potentially equivalent residues and local directions of the main chain. The superposition and updating is repeated until no increase can be obtained in the number of equivalences.

In general, the resulting superimposed 3D structures show that large parts of the two proteins are very similar in structure and hence appear to be the structurally conserved regions, while other sections differ considerably. It should be noted that the applied algorithms do not take into account explicitly the secondary structure. Since—according to the definition SCRs must be terminated at the end of a secondary structural unit, so that, for example, each single strand of a β-sheet comprises a separate SCR—secondary structural elements of the proteins must be assigned before SCRs are determined. Information about the secondary structure of any known protein can be derived in the easiest way from crystal data files (for example from the PDB files) which include the secondary structural elements detected by crystallographers. Because the assignment of secondary structures in crystal structure files is often subjective and sometimes incomplete, it is more convenient to use objective methods which are able to assign correctly the secondary structural elements. Programs like DSSP [41] or STRIDE [42] detect secondary structural elements on the basis of geometrical features, i.e. the hydrogen bonding pattern or the main chain dihedral angle. Using these programs—which are available free of charge from the EMBL server in Heidelberg—one can rapidly assign secondary structures to all proteins if atomic coordinates exist.

The situation is more complicated when only a single homologous protein is known that can be used as reference structure for the target sequence, because with only one known template protein a basis for a structural comparison does not exist. Under these circumstances one has to detect the SCRs manually using both, sequence and structural information of the proteins. As was described before, conserved regions are frequently detected in stable secondary structure elements. Therefore, it is reasonable to study carefully as many of those elements as possible in the reference protein with the aim of discovering potential clues for the existence of SCRs. Residues in the hydrophobic core tend to be more conserved with regard to sequence and 3D structure than residues at the protein surface. Amino acids involved in salt bridges, hydrogen bonds and disulfide bridges are most likely to be conserved within a protein family. The same holds true for amino acids located in the active site. Information derived from multiple sequence alignments can also be used beneficially to locate the SCRs more accurately.

It was found in many investigations on homologous proteins, that SCRs show strong sequence homology, while the variable regions show little or no sequence homology and are the sites of addition and deletion of residues. For that reason the determined SCRs should have identical or closely homologous sequences. Due to the structural homology of these regions no gaps are allowed in conserved areas.

In cases where the SCRs of the reference proteins already are known one has only to locate the regions of the model protein that correspond to these SCRs. This is accomplished by aligning the target sequence with the sequences of the SCRs in the homologs. The alignment procedure which must be applied for this purpose differs slightly from that already described. Because, by definition, SCRs cannot contain insertions or deletions, an algorithm is needed which disallows the introduction of gaps within SCRs. Unfortunately the standard Needleman and Wunsch method does not have the measure for treating SCRs in a special manner. It places a gap at any location if this results in an optimized amino acid matching. For this reason procedures have been developed [3, 22, 43] which can handle each SCR independently. Corresponding programs generate alignments without gaps appearing within any conserved region. When the correspondence between the reference and the target sequences has been established the coordinates for the SCRs can be assigned. The coordinates of the reference proteins are used as basis for this assignment. In segments with identical side chains detected in reference and target proteins, all coordinates of the amino acids are transferred. In diverse regions only the backbone coordinates are transferred. The corresponding side chains then will be added after complete backbone (SCRs and SVRs) generation (see section 4.3.4).

4.3.3 Construction of Structurally Variable Regions (SVRs)

Since significant differences in protein structures occur preferably in loop regions, the construction of these *structurally variable regions* (SVRs) is a more challenging task. Insertions and deletions due to differences in the number of amino acids additionally complicate the modeling procedure. A variety of methods for generating loops have been developed and described comprehensively in the literature [5–7, 44–46]. A good guide for modeling the missing region can be the structure of a segment of equivalent length in a homologous protein. Extensive investigations of variable regions in homologous proteins have shown that in cases where particular loops possess the same length and amino acid character, their conformation will be the same. The coordinates then can be transferred directly to the model protein in the same way as described for the SCRs. If no comparable loop exists in the protein family, two other strategies can be applied for modeling the SVRs. The coordinates for the SVRs can be either retrieved from peptide segments which are found in other proteins and that fit properly into the model's spatial environment [5–7], or by generating a loop segment de novo [44–46]. The former approach, which is known as *loop search method*, looks for peptide segments in proteins which meet the specified geometrical criterion. Usually the loop search programs are scanning the Brookhaven protein data bank for possible peptide segments. The specified geometry input for the database search is given by distances and coordinates, including the residues of the regions embracing the loop segment in the model. The output of a respective search is a collection of loops satisfying the specific geometrical constraints. Usually the 10 to 20 best loop fragments are retained for further examination. The loops are ranked according to goodness of fit to the desired structure. However, additional criteria not used explicitly during the loop search, can provide a guide to ascertain the preference of one loop candidate over another. The retrieved fragments can be analyzed on the basis of quality of fit to the residues confining the loop region, by determining sequence homology between the original loop sequence and the sequence of the retrieved fragment, or via evaluation of steric interactions and energy criteria.

The loop search method offers the advantage that all loops found are guaranteed to possess reasonable geometry and resemble known protein conformations. It is not certain that the chosen segment fits properly into the existing framework of the model, so severe sterical overlaps may be detected. If this happens, the *de novo generation technique* is an alternative method.

Using this approach a peptide backbone chain is built between two conserved segments using randomly generated numerical values for all the backbone dihedral angles. Several algorithms have been developed to optimize the search strategy and to reduce computing time. Due to the complexity of this type of search method the approach can only be used for loops smaller than seven residues.

All loops generated by database or random search methods are usually far from optimal geometry. For that reason all loop regions (including confining residues) must subsequently be refined by energy minimization techniques in order to remove steric hindrance and to relax the loop conformations (see section 4.4.3).

4.3.4 Side Chain Modeling

When the peptide backbone has been constructed the next step is to add the side chains. The prediction of the numerous side chain conformations is by far a more complex problem than the prediction of the backbone conformation of a homologous protein. Many of the side chains possess one or more degrees of freedom and therefore can adopt a variety of energetically allowed conformations.

Several strategies have been developed in the past to find a solution for this multiple minima problem [47–54]. It has been generally assumed that identical residues in homologous proteins adopt similar conformations. Also, when the substituted side chain belongs to an amino acid pair that shows high homology (indicated by a high score in the Dayhoff matrix, for example Ile and Val, or Gln and Glu), it is assumed that the side chains adopt the same orientation in the protein [47]. The situation will become more complicated if the amino acids to be substituted are not related. When the side chain to be considered is longer than its counterpart in the homologous protein or is structurally dissimilar, the side chains must be positioned at random but in a conformation that avoids unfavorable contacts with other side chains [48]. An alternative way to obtain a suitable side chain conformation is to select the calculated minimum conformations of the appropriate dipeptide potential energy surface [49].

A more reliable procedure was developed by examining the relationship between the side chain positions in homologous structures of globular proteins. It has been found that the side chains adopt usually only a small number of the many possible conformations [50, 51]. Side chains with for example two χ angles have been observed to exist in four to six common conformations. All observed rotamers are combinations of the familiar gauche and anti forms. On the basis of such statistical evaluations rotamer libraries have been developed [50, 53]. The most often applied side chain library is the one created by Ponder and Richards [50] which contains 67 rotamers for 17 amino acids. Several homology modeling programs make use of this library for generating the side chains of homologous proteins. Selecting the most probable conformation out of a rotamer library for side chain modeling might be problematical because this procedure disregards the information that is available from the equivalent side chain of the reference structure. Apart from that, the correct conformation of a side chain depends essentially on the local environment met by the amino acid in the real protein. This has been shown by several authors who have investigated well-resolved protein structures [54, 55]. In the interior of a protein, hydrophobic interactions are predominant and result in tight packing of amino acid residues. Factors such as the secondary structure and tertiary contacts with other residues can influence the side chain conformation. For that reason, methods have been developed which take into account information about the local environment and other constraints which may determine the positions of side chains. Sutcliffe et al., for example, have developed rules for mutual substitution of all 20 naturally occurring side chains in α-helical, β-sheet and loop regions—a total of $20 \times 20 \times 3 = 1200$ rules [54]. In order to determine which atom positions are preserved when substituting one amino acid for another at a topologically equivalent position, the study was performed on several sets of homologous proteins. All residues corresponding to a particular topologically equivalent position were aligned on their backbone atoms and inspected to determine which atoms are correlated in spatial position.

As we have discussed, various methods for the modeling of side chains do exist. All of them can greatly assist the modeler by providing appropriate side chain conformations. On the other hand, in several situations one has to refine side chain positions manually. Modifications must be applied, for example, when amino acids are involved in specific interactions like ion-pair formations, disulfide bridges, buried charge interactions or internal hydrogen bonds. Variations also occur when the residues are located on the protein surface and are fully accessible. Such exceptions must be treated on a case-by-case basis.

Once the final model has been built, a refinement of the structure is usually desirable. Regions where SCRs and SVRs are connected usually suffer from high steric strain and must be minimized. Several side chains may also adopt positions which result in bad van der Waals contacts. A stepwise approach for the structure refinement is likely to produce the best result. Overall simultaneous optimization of all side chains possibly would destroy important internal hydrogen bonds and may cause a conformational change within conserved regions. In order to remove steric overlaps, conformational searches are applied for residues which show bad van der Waals contacts. Energy minimizing and/or molecular dynamics of the model are useful routes to explore the local region of conformational space and may produce a more refined structure. The details about energy minimization and molecular dynamics used for structure refinement will be described comprehensively in section 4.4.3.

4.3.5 Distance Geometry Approach

While several reference structures are often used in the traditional homology modeling process, only one set of coordinates can be used for the construction of a particular structurally conserved region (see section 4.3.2). The distance geometry approach in homology modeling [38, 56, 57] offers the possibility to examine all the reference proteins simultaneously to impose structural constraints that in turn can be used to generate conformations consistent with the data set. The first step using this procedure is the same as in the traditional homology modeling approach. The SCRs are identified and the sequence of the target protein is aligned with the sequences of the known proteins. The distance geometry method applies rules by which a multiple sequence alignment can be translated into distance and chirality constraints, which are then used as input for the calculation. By this means one obtains an ensemble of conformations for the unknown structure, where each member of the ensemble contains regions (which were constrained during the calculation) showing similar conformations and regions (which were free during the calculation) with varying arrangements. The structures of the ensemble then are energy-minimized in order to eliminate structural irregularities that sometimes appear during distance geometry calculations. The differences among the derived conformations provide an indication of the reliability of the structure prediction. A detailed description of this technique is given in a study reporting the application of the method to predict the structure of flavodoxin from *E. coli* [58].

4.3.6 Secondary Structure Prediction

The best method for the generation of a structure proposal for a protein with unknown 3D structure is to base it on a homologous protein whose 3D structure *is* available, i.e. by means of the knowledge-based approach as described earlier. However, in cases where a homologous protein does not exist, several other methods have been developed that have concentrated on the prediction of secondary structure. The underlying idea evolves from the fact that 90% of the residues in most proteins are engaged either in α-helices, β-strands or reverse turns. As a consequence it seems possible—if the secondary stuctural elements are predicted accurately— to combine the predicted segments in an effort to generate the complete protein structure. Obviously the reliability of this approach is much lower than homology modeling, thus, it should be applied with extreme caution. However, the prediction of secondary structure from the amino acid sequence has been widely practiced (for reviews see [59–67]).

Basically, three different types of methods can be employed for this task: statistical, stereochemical and homology/neural network-based methods. All different prediction methods rely, more or less, on information derived from known 3D structures stored in the Brookhaven protein database. The correct assignment of secondary structural regions in the crystal structure (see section 4.3.2) is therefore necessary for a reliable validation of all prediction methods.

Statistically based methods were among the first that have been developed. The underlying idea takes advantage of the observation that many of the 20 amino acids show statistically significant preferences for particular secondary structures. Ala, Arg, Gln, Glu, Met, Leu and Lys for example are preferentially found in α-helices, whereas Cys, Ile, Phe, Thr, Trp, Tyr and Val occur more frequently in β-sheets. The most simple and most commonly used statistical method for secondary structure prediction is the one proposed by Chou and Fasman [60]. The prediction is done by calculating the probability of an amino acid to belong to a particular type of secondary structure, such as α-helix, β-sheet or turn, based simply on its frequency of occurrence as part of the respective secondary structure elements as found in the Brookhaven protein database. Another commonly used statistical-based method is that of Garnier, Osguthorpe and Robson (GOR) [61]. The success of this type of algorithm is difficult to verify because some of them merely produce tendencies towards a particular secondary structure rather than an absolute prediction. Therefore, the methods are open to divergent interpretations, with the result that different authors obtain different results. The scope and limitations of the statistical methods have been demonstrated by Kabsch and Sander [68] in an analysis of three commonly used prediction methods showing that all methods are below 56% accurate in predicting helix, sheet and loop.

Another type of secondary structure prediction method is based on the interpretation of the hydrophobic, hydrophilic and electrostatic properties of side chains in terms of the formulation of rules for the folding of proteins [63–65]. The method of Lim, for example, takes into account the interactions between side chains separated by up to three residues in the sequence [63] in view of their packing behavior in either the α-helical or β-sheet conformations. A sequence with alternating hydrophobic and hydrophilic side chains, for example, is likely to be found in a β-sheet strand, with hydrophilic residues exposed to the solvent and hydrophobic residues buried in the interior of the protein. Correspondingly, the stereochemical-based methods have been applied successfully for the prediction of amphiphilic helices [64] or membrane-spanning segments [65].

Other procedures combine statistical and stereochemical rules in a single algorithm to predict the secondary structure. Examples are the JAMSEK [69] and the ALB [70] programs.

Recently, Sander has reported an algorithm which uses evolutionary information contained in multiple sequence alignments as input to neural networks [66, 67]. Neural networks potentially have a methodological advantage compared with other prediction methods because they can be trained. This means that rules determining the behavior of the studied systems are not needed in advance, but are formed by the network itself on the basis of known facts. In a recently published study, a neural network method (called PHD) showed 70% accuracy in the prediction of three classes of secondary structure (helix, sheet, loop) on the basis of only one known homologous sequence [67].

Information derived from secondary structure prediction of homologous proteins is often used in addition to the results received in a primary sequence alignment in order to improve the location of the SCRs in a class of homologous proteins. Even when only the structure of one homologous protein is known (which can be used as template for the homology modeling approach), but several homologous sequences, it is helpful to include the predicted secondary structural elements for the homologous sequences to assign the SCRs. All available prediction methods should be applied in order to find the most probable assignment for the secondary structural elements. Of course different methods do not yield exactly the same result. This is shown in Fig. 3 using five methods (CHOU, GOR, ALB, JAMSEK, PHD) for the prediction of the known secondary structure of a cephalosporinase from *Enterobacter cloacae*. The prediction is also compared with the result of the DSSP program, which assigns the secondary structure on the basis of the known atomic coordinates.

Most of the prediction methods described are implemented in commercially available protein modeling programs. However, nowadays the World Wide Web can also be used for structure prediction purposes. At the EMBL in Heidelberg (http://www.embl-heidelberg.de), for example, an automated mail server is installed which offers a variety of secondary structure prediction methods, including a neural network-based method.

4.3.7 Energy-Based Modeling Methods

In contrast to the knowledge-based techniques which, as described, check a database of known structures of homologous proteins for the most probable conformation of a specified region, energy-based methods generate a protein model ab initio, founded solely on the primary sequence. Whereas modeling with knowledge-based procedures is based on a set of empirically and statistically proven rules, energy-based methods use a list of geometrical criteria in order to sample all possible conformations of a defined region and to find the conformation of lowest energy. Energy-based approaches can be regarded as approximate solutions for the protein folding problem. Various programs based on the energy approach are available: the SCEF (self-consistent electrostatic field) techniques [71], Monte Carlo methods [72] and procedures which make use of empirically derived force fields (knowledge-based force fields) [73, 74].

The so-called mean force potentials, or knowledge-based force fields, are quite different from the traditional force fields (which were described in general in section 2.2.1). The basic idea of knowledge-based force fields is that molecular structures observed from X-ray analysis or NMR contain a wealth of information on the stabilizing forces within macromolecules. Using statistical methods, the underlying rules governing the 3D structure of proteins have been revealed. It is the basic assumption of the Boltzmann principle that frequently observed states correspond to low-energy states of a system. Thus, the mean force potentials are compiled by extracting relative frequencies of particular atom pair interactions from a database of protein structures [75]. The mean force potentials consist usually of interactions among particular atom pairs and protein–solvent interactions. They incorporate all kinds of forces (electrostatic, dispersion, etc.) acting between particular protein atoms as well as the influence of the surrounding solvent on the interaction and can therefore be used to predict the structure of a macromolecule from its primary sequence. Mean force potentials have been applied for the prediction of protein folds and even for the detection of errors in protein models and experimentally determined structures [76].

```
                   1                                                50
SEQUENCE           MMRKSLCCALLLGISCSALATPVSEKQLAEVVANTITPLMKAQSVPGMAV

CHOU                     EEEEEEEEE    HHHHHHHHHHHHH   HHHHHHHHHHTTEEE
GOR                H  HH     E                 HHHHHHHHH HH  HH        EEE
ALB                    HHHHHHHHHHTTTTT TTT    HHHHHHHHHHHHHHH    TTEEE
JAMSEK             HHHHHHHH        TTT      HHHHHHHHTTT                 E
PHD                TTTT HHHHHHHHHHHHH    T   HHHHHHHHHHHHHHHH    TTTTT EE
DSSP                               HHHHHHHHHHHHHHHHHHHH     EEEEEEE


                   51                                               100
SEQUENCE           AVIYQGKPHYYTFGKADIAANKPVTPQTLFELGSISKTFTGVLGGDAIAR

CHOU               EEEEE     EEEEE HHHHH    EEEEEEE         EEEEEETTTHHHH
GOR                EEEE       EEE HHHHH        EHEEHE H    EEEEEE      EH
ALB                EEEE TTTTEEEE         TT TT     HHHHHHHHHHHHHTTHHHHH
JAMSEK             EEEE TTT EEEETTT     TTT    TTT        TTTEEEE   HHHHHH
PHD                EEEE TT EEEE      TTTTTTTTT                         H
DSSP               ETTEEEEEEEEEEEETTTT   EE TTTTEEEE       HHHHHHHHHHH


                   101                                              150
SEQUENCE           GEISLDDAVTRYWPQLTGKQWQGIRMLDLATYTAGGLPLQVPDEVTDNAS

CHOU               HHHHHHHEEEEEE    TT      HHHHHHHH  TT EEEEETT    TTTTT
GOR                 HE   HHHEE E HH      HHHEEHHHH  H               HHH HH
ALB                 THHHHHHHHH              TTTEEEEETTT   TTHH
JAMSEK             HH HHHHHHHHH  TTT  TTTHHHHHHHHHEEE         TTT   TTHH
PHD                    TTT    TT H       T      HHHHH TTTTT TT   H HHH
DSSP               TTTT      TTTTTT    TTT HHHHHH  TTTTTTTTTTTT  HHH


                   151                                              200
SEQUENCE           LLRFYQNWQPQWKPGTTRLYANASIGLFGALAVKPSGMPYEQAMTTRVLK

CHOU               EEEEEEEEETT   TTE EEEEEEEEEEEEE HHHHHHTTT     HHHHHHHHHH
GOR                HHEEHH              EEEHH   HHHHHH         HHHHHHHHH
ALB                HHHHH     TTT  TTT       HHHHHHHH  TTT HHHHHHHHHHH
JAMSEK             HHHHHHTTTTTTTT    EEETTT  EEEEE    TTT          HHHHHH
PHD                HHHHHHH TTTTTTTT EE   TT    HHHHHH    TTT HHHHHHHHHHH
DSSP               HHHHHHH       TTTTEE    HHHHHHHHHHHHH    HHHHHHHH


                   201                                              250
SEQUENCE           PLKLDHTWINVPKAEEAHYAWGYRDGKAVRVSPGMLDAQAYGVKTNVQDM

CHOU               HHHHHH EEEEEHHHHHHHH          TTTHHHHHHHTTHHHHHHH    EEEEEEHH
GOR                 H H HHHHHH  HHHHHHHHHH H       EEE   HHHHHHH   E HHHH
ALB                    EEEEETTTT              TTT EEE HHHHHHHHHHHHHHHHHHHH
JAMSEK             HHHTTTEEEEE          TTT   EEE                EEEEE
PHD                H TTTT    TTHHHHHHHHHH   TTTT EE TTT T     TTT HHHHH
DSSP                   TTTEETTT            EETTTTEE    TTTHHHHHHEEEEHHHH
```

```
              251                                                        300
SEQUENCE      ANWVMANMAPENVADASLKQGIALAQSRYWRIGSMYQGLGWEMLNWPVEA

CHOU          HHHHHHHHHHHHHHHHHHHH  EEEEEEEEEE   EEEEE HHHHHHHHHH
GOR           HHHHHH H   H HHHHHHHHHHHHHH   EEEEEEE        HHHE
ALB           HHHHHHHH HHHHHHHHHHHHHHHHHHHHHHH       HHHHHHHHHHHHH
JAMSEK          EEE     HHHHHHHHHHHHHHEEEETTTEEEEEETTT                 T
PHD           HHHHHH  T  T  HHHHHHHHHHH                T       TTTTT
DSSP          HHHHHHHH        HHHHHHHHHHH EEEEETTEEEETTTTEEEETTT  H


              301                                                        350
SEQUENCE      NTVVEGSDSKVALAPLPVAEVNPPAPPVKASWVHKTGSTGGFGSYVAFIP

CHOU          HHHHH TTHHHHHHHHHHHHHH TT    HHHHHH TTTTTTTEEEEEEEE
GOR               EEE      HHH     H          H  EEEEE    E EEEEE
ALB           EEEETTTTEEEEE EEEEEE TT                       EEEEET
JAMSEK        TT      TTT         TTT                        EEEE
PHD               TTT      TTTTT    TTTTT        E   TT  T   EEEEE
DSSP          HHHHHHH HHHHH  EE EEEEEE  TTTEEEEEEEEEETTEEEEEEEE


              351                              381
SEQUENCE      EKQIGIVMLANTSYPNPARVEAAYHILEALQ

CHOU             EEEEEEEEEETTTTTTTHHHHHHHHHHHHHH
GOR           HHHHHEEEEE            HHHHHHHHHHHHHH
ALB           TTEEEEEEE            TT HHHHHHHHHHH
JAMSEK          EEEEE    TTT
PHD              EEEEE   TTTTTHHHHHHHHHHHHHHH  T
DSSP             EEEEEEE       HHHHHHHHHHHHHH
```

Figure 3: Comparison of secondary structure predictions using different methods for a crystallographically resolved cephalosporinase from *Enterobacter cloacae*. Sturcture elements shown in real agree with the structures observed in the crystal (H = α-helix, E = β-strand, T = turn)

The SCEF method is based on the idea that the electrostatic interaction is important for the tertiary structure of a protein. In this procedure an initial approximation is made by calculating only the electrostatic energy and assuming that each amino acid must have optimal electrostatic energy, i.e. the dipole moment of each residue must be optimally aligned in the electrostatic field generated by the whole protein. As long as this is not achieved, the orientation of the dipole moment of each amino acid is changed to improve its electrostatic energy. Then the energy of the whole protein, including all energy terms, is minimized and the procedure is repeated iteratively to achieve self-consistency. This method has been tested successfully on several peptides [71].

Energy-based approaches are still under development. It is true that they have already been applied successfully in predicting the general folds of some proteins where no information on the secondary or teriary structure was available, but further improvement is necessary before these techniques can be employed as standard procedures for the prediction of the complete tertiary structure of any unknown protein.

References

[1] Blundell, T. L., and Johnson, L. N. *Protein Crystallography*, Academic Press: New York 1976.

[2] Bashford, D., Chothia, C., and Lesk, A. M. *J. Mol. Biol.* **196**, 199–216 (1987).

[3] Greer, J. *J. Mol. Biol.* **153**, 1027–1042 (1981).

[4] Chothia, C., and Lesk., A. M. *J. Mol. Biol.* **160**, 309–342 (1982).

[5] Johnson, M. S., Srinivasan, N., Sowdhamini, R., and Blundell, T. L. *Crit. Rev. Biochem. Mol. Biol.* **29**, 193–316 (1994).

[6] Sali, A., Overington, J. P., Johnson, M. S., and Blundell, T. L. *TIBS* **15**, 235–240 (1990).

[7] Jones, T. A., and Thirup, S. *EMBO J.* **5**, 819–822 (1986).

[8] Dudek, M. J., and Scheraga, H. A. J. *Comput. Chem.* **11**, 121–151 (1990).

[9] Levin, R. *Science* **237**, 1570 (1987).

[10] Thornton, J. M., and Gardner, S. P. *Trends Biochem. Sci.* **14**, 300–304 (1989).

[11] Orengo, C. A., Brown, N. P., and Taylor, W. R. *Proteins* **14** 139–146 (1992).

[12] Lipman, D. J., and Pearson, W. R., *Science* **227**, 1435–1441 (1985).

[13] Altschul, S. F., Gish, W., Miller, W., Myers, E. W., and Lipman, D. J. *J. Mol. Biol.* **215**, 403-410 (1990).

[14] HOMOLOGY User Guide, Biosym Technologies, San Diego, USA.

[15] MODELLER, Molecular Simulations Inc., Burlington, Massachusetts, USA

[16] SYBYL COMPOSER Theory Manual, Tripos Associates, St. Louis, Missouri, U.S.A.

[17] WHAT IF, Vriend, G., European Molecular Biology Laboratory (EMBL), Heidelberg, Germany.

[18] Devereux, J., Haeberli, P., and Smithies, O. *Nucleic Acids Res.* **12**, 387–395 (1984).

[19] Barton, G. J. *Methods Enzymol.* **183**, 403–428 (1990).

[20] Needleman, S. B., and Wunsch, C. D. *J. Mol. Biol.* **48**, 443–453 (1970).

[21] Sander, C., Schneider, R. *Proteins* **9**, 56–58 (1991).

[22] Schuler, G. D., Altschul, S. F., and Lipman, D. J. *Proteins Struct. Func Gen.* **9**, 180–190 (1991).

[23] Vingron, M., and Argos, P. *Comput. Appl. Biosci.* **5**, 115–121 (1989).

[24] Bowsell, D. R., and McLachlan, A. D. *Nucleic Acids Res.* **12**, 457–465 (1984).

[25] Dayhoff, M. O., Schwartz, R. M., and Orcutt, B. C. A Model of Evolutionary Change in Proteins. In: *Atlas of Protein Sequence and Structure*, Vol. 5, Suppl. 3. Dayhoff, M. O. (Ed.). Natl. Biomed. Res. Found.: Washington; 345–352 (1978).

[26] Gribskov, M., McLachlan, A. D., and Eisenberg, D. *Proc. Natl. Acad. Sci. U.S.A.* **84**, 4355–4358 (1987).

[27] Risler, J. L., Delorme, M. O., Delacroix, H., and Henaut, A. *J. Mol. Biol.* **204**, 1019–1029 (1988).

[28] Engelman, D. M., Steitz, T. A., and Goldman, A. *Anu. Rev. Biophys. Chem.* **15**, 321 (1986).

[29] Gonnet, G. H., Cohen, M. A., and Benner, S. A. *Science* **256**, 1443–1445 (1992).

[30] Doolittle, R. *Methods Enzymol.* **183,** 736–772 (1990).

[31] Barton G. J., and Sternberg, M. J. E. *J. Mol. Biol.* **198**, 327–337 (1987).

[32] Chothia, C., and Lesk, A. M. *EMBO J.* **5**, 823–826 (1986).

[33] Perutz, M. F., Kendrew, J. C., and Watson, H. C. *J. Mol. Biol.* **13**, 669–678 (1965).

[34] Kabsch, W. *Acta. Cryst.* **A32**, 922–923 (1976).

[35] Kabsch, W. *Acta. Cryst.* **A34**, 827–838 (1978).

[36] McLachlan, A. D. *Acta. Cryst.* **A38**, 871–873 (1982).

[37] Taylor, W. R. *J. Mol. Biol.* **188**, 233–258 (1986).

[38] Crippen, G. M., and Havel, T. F. *Acta. Cryst.* **A34**, 282–284 (1978).

[39] Vriend, G., and Sander, C. *Proteins* **11**, 52–58 (1991).

[40] Matthews, B. W., and Rossmann, M. G. *Methods Enzymol.* **115**, 397–420 (1985).

[41] Kabsch, W., and Sander, C. *Biopolymers* **22**, 2577–2637 (1983).

[42] Frishman, D., and Argos, P. *Proteins Struct. Func. Gen.*, in press (1995).

[43] Fredman, M. *Bull. Mathematical Biology* **46**, 553–558 (1984).

[44] Bruccoleri, R. E., and Karplus, M. *Biopolymers* **26**, 137–168 (1987).

[45] Bruccoleri, R. E., Haber, E., and Novotny, J. *Nature* **335**, 564–568 (1988).

[46] Shenkin, P. S., Yarmush, D. L., Fine, R. M., Wang, H., and Levinthal, C. *Biopolymers* **26**, 2053–2085 (1988).

[47] Feldmann, R. J., Bing, D. H., Potter, M., Mainhart, C., Furie, B., Furie, B. C., and Caporale, L. H. *Ann N. Y. Acad. Sci.* **439**, 12–43 (1985).
[48] Blundell, T. L., Sibanda, B. L., and Pearl, L. *Nature* **304**, 273–275 (1983).
[49] Palmer, K. A., Scheraga, H. A., Riordan, J. F., and Vallee, B. L. *Proc. Natl. Acad. Sci. U.S.A.* **83**, 1965–1969 (1986).
[50] Ponder, J., and Richards, F. M. *J. Mol. Biol.* **193**, 775–791 (1987).
[51] Summers, N. L., Carlson, W. D., and Karplus, M. *J. Mol. Biol.* **196**, 175–198 (1987).
[52] McGregor, M. J., Islam, S. A., and Sternberg, M. J. *J. Mol. Biol.* **198**, 195–210 (1987).
[53] Benedetti, E., Morelli, G., Nemethy, G., and Scheraga, H. A. *Int. J. Peptide Protein Res.* **22**, 1–15 (1983).
[54] Sutcliffe, M. J., Hayes, F. R. F., and Blundell, T. L. *Protein Eng.* **1**, 385–392 (1987).
[55] Schrauber, H., Eisenhaber, F., and Argos, P. *J. Mol. Biol.* **230**, 592–612 (1993).
[56] Havel, T. F., and Snow, M. *J. Mol. Biol.* **217**, 1–7 (1991).
[57] Srinivasan, S., March., C. J., and Sudarsanam, S. *Protein Science* **2**, 277–289 (1993).
[58] Havel, T. F. *Molecular Simulations* **10**, 175–210 (1993).
[59] Fasman, G. D. *Trends Biochem. Sci.* **14**, 295–299 (1989).
[60] Chou, P. Y., and Fasman, G. D. *Biochemistry* **13**, 211–245 (1974).
[61] Garnier, J., Osguthorpe, D. J., and Robson, B. *J. Mol. Biol.* **120**, 97–120 (1978).
[62] Biou, V., Gibrat, J. F., Levin, J. M., Robson, B., and Garnier, J. *Protein Eng.* **2**, 185–191 (1988).
[63] Lim, V. I. *J. Mol. Biol.* **88**, 873–894 (1974).
[64] Eisenberg, D., Weiss, R. M., and Terwilliger, T. C. *Nature* **299**, 371–374 (1982).
[65] Kyte, J., and Doolittle, R. F. *J. Mol. Biol.* **157**, 105–132 (1982).
[66] Rost, B., and Sander, C. *J. Mol. Biol.* **232**, 584–599 (1993).
[67] Rost. B., and Sander, C. *Proteins* **19**, 55–72 (1994).
[68] Kabsch, W., and Sander, C. *FEBS Lett.* **155**, 179–182 (1983).
[69] Mrazek, J., and Kypr, J. *Comput. Appl. Biosci.* **4**, 297–302 (1988).
[70] Ptitsyn, O. B., and Finkelstein, A. V. *Biopolymers* **22**, 15–25 (1983).
[71] Piela, C., and Scheraga, H. A. *Biopolymers* **26**, S33–S58 (1987).
[72] Skolnick, J., and Kolinski, A. *Anu. Rev. Phys. Chem.* **40**, 207–235 (1989).
[73] Sippl, M. J. *J. Mol. Biol.* **213**, 859–883 (1990).
[74] Bowie, J. U., Luethy, R., and Eisenberg, D. *Science* **253**, 164–169 (1991).
[75] Sippl, M. J. *J. Comput.-Aided Mol. Design* **7**, 473–501 (1993).
[76] Sippl, M. J. *Proteins* **17**, 355–362 (1993).

4.4 Optimization Procedures—Model Refinement— Molecular Dynamics

4.4.1 Force Fields for Protein Modeling

Protein models derived from either homology modeling and alignment studies or crystal structures need further refinement. In the course of generating protein models the loop and side chain conformations in general are chosen arbitrarily; therefore the conformations do not correspond to energetically reasonable structures. Also crystal structures must be relaxed in order to remove the internal strain resulting from the crystal packing forces or to remove close contacts between hydrogen atoms or amino acid residues which may have been added to the crystal coordinates after structure determination.

As protein models consist of hundreds or thousands of atoms the only feasible methods of computing systems of such size are molecular mechanics calculations. The common force fields used in molecular mechanics calculations are based in principle on the equations for the potential energy function as described in section 2.2.1. However, force fields for protein modeling differ in some respect from small molecule force fields. Besides the specific parametrization for proteins and DNA, certain simplifications are frequently introduced. In some force fields non-polar hydrogens are not represented explicitly, but are included into the description of the heavy atoms to which they are bonded. In contrast, polar hydrogens which may act as potential partners in hydrogen bonding are treated explicitly. This procedure is denoted as the *united atom model*. In the AMBER [1, 2] force field both the united-atom model or an all-atom representation can be applied, while the GROMOS force field [3] offers only the united atom model. Other simplifications can be made by introducing cut-off radii [4] to reduce the time-consuming part of calculating non-bonded interactions between atoms separated by distances larger than a defined cut-off value.

An additional variation is made in respect of the treatment of the electrostatic interactions. As the explicit inclusion of solvent is still a problem, some force fields try to simulate the solvent effect by introducing a distance-dependent dielectric constant [1, 2]. Especially in the case of macromolecules the electrostatic field in the environment of the system can not be considered to be continous. Thus, a differentiating procedure in calculating the particular properties is necessary in order to reflect the electrostatic effects which depend on the local situation, e.g. in the binding pocket or on the surface of the protein. A detailed discussion of this subject and a description of methods handling the complex situation adequately is given in section 4.6.1.

The modifications established in protein force fields are various and can not be discussed here in detail. A comprehensive description of potential simplifications is given in [5]. It should be borne in mind that each simplification applied can result in a loss of accuracy. The decision on the force field to be chosen strongly depends on the problem to be investigated; hence the most accurate force field which is applicable for the whole study must always be selected. The use of different force fields within a molecular modeling study should generally be avoided.

There are several common force fields for protein modeling implemented in software programs. The following list is not complete but comprises some of the most frequently employed methods: AMBER [1, 2], CVFF [6], CHARMM [7] and GROMOS [3].

4.4.2 Geometry Optimization

The algorithms used in the minimization procedures for proteins are the same as for small molecules and have been discussed in detail in section 2.2.3. The minimization algorithms applied to optimize the geometry usually find only the local minimum on the potential energy surface closest to the initial coordinates. In case of a well-resolved crystal structure the minimization will directly yield one energetically favorable conformation. The relaxation of a crystal structure usually is a straightforward procedure. However, crystal coordinates—even if highly resolved—sometimes have several unfavorable atomic interactions. These disordered atomic positions cause large initial forces that result in artificial movements away from the original structure when starting the minimization process. A general approach to avoid these large deviations is to relax the protein model gradually.

A more profound solution would be to assign tethering forces to all heavy atoms of the crystal structure in the first stage of minimization. The *tethering constant* is a force applied to fix atomic coordinates on predefined positions. The strength of the tethering force can be selected by the user and affects the extent of movement of the atoms measured by the rms deviation from the initial coordinates. When tethering the heavy atoms the hydrogen atoms, and perhaps solvent molecules, are allowed to adjust their positions in order to minimize the total potential energy. A suitable minimization method for this purpose is the steepest descent algorithm. For this initial relaxation step a crude convergence criterion can be applied or the process can be finished by defining a maximum number of allowed minimization steps.

Subsequently it is recommended to tether only the well-defined main chain atoms. Now the side chains are allowed to move and to adjust their orientations. The steepest descent method is suitable also in this case. Ultimately the restraints are removed in the last step so that the final minimum represents a totally relaxed conformation. The minimization algorithm should be changed to conjugate gradient to reach convergence in an effective way. The convergence criterion should be fixed on the order of 0.002 kcal mol^{-1} Å^{-1} to ensure a final geometry nearest to the minimum.

The application of tethering forces can also be useful and necessary in the modeling of incomplete systems. These may result in an X-ray study if certain parts of the crystals or included solvent molecules cannot be resolved adequately. Also active site models of enzymes or binding pockets of proteins used for the investigation of potential ligand–protein interactions are examples of typical incomplete systems.

Due to the absence of neighboring amino acids or solvent molecules the atom positions at the surface of a protein are mobile. As a consequence, large deviations from the initial positions will result after minimization and the final geometry must be regarded as an artefact. Therefore atoms or the ends of side chains are tethered at their original positions to avoid unrealistic atom movements at the surface of the protein.

With the objective to confirm the accuracy of the relaxed protein model the deviations from the experimental structure should be examined. For this purpose the initial structure and the final geometry are superimposed using least-squares fit methods. Normally either all backbone atoms or only backbone atoms of the well-refined secondary structural elements are used as fitting points. The quality of the fit can be judged by the rms deviation of the optimized form from the initial geometry. The value of the rms deviation is strongly dependent on the number and localization of atoms which are considered for the fit. Naturally, a fit of all heavy atoms would result in a much higher rms value than a fit which is confined to backbone atoms only, mainly due to the greater mobility of side chains.

If the generated model is based merely on homology modeling and alignment studies the loop and side chain conformations need further refinement. It is necessary to investigate carefully their conformational behavior and to analyze the potential energy surface for other possible low-energy conformations. A valuable tool for this purpose are *molecular dynamics simulations*. The relaxed geometry obtained as result of a minimization procedure can be used as starting point for molecular dynamics simulations.

4.4.3 The Use of Molecular Dynamics Simulations in Model Refinement

As mentioned above the refinement of models derived from homology modeling and alignment studies is a must. Loop and side chain conformations of the derived protein model represent only one possible conformation and the minimum structure found by the minimization algorithms represents only one local minimum. In order to detect the energetically most favored 3D structure of a system a modified strategy is needed for searching the conformational space more thoroughly.

Molecular dynamics simulations offer an effective means to solve this problem, especially for molecules containing hundreds of rotatable bonds. A molecular dynamics simulation is performed by integrating the classical equations of motion over a period of time for the molecular system. The resulting trajectory for the molecule can be used to compute the average and time-dependent properties of the system. The theory of the molecular dynamics method and its application in conformational searching of small molecules have been discussed and illustrated on some impressive examples in section 2.3.3. Here, we will focus on the utilization of this technique in the refinement of 3D macromolecular structures.

The use of molecular dynamics has made an essential contribution to the understanding of dynamic processes in proteins at the atomic level. However, there are some basic limitations and problems arising with increasing size and associated with the immense number of degrees of freedom of large molecular systems.

Although computer resources have become sufficiently powerful to enable handling of quite large systems (up to 50000 atoms) it is still necessary to introduce some modifications in order to reduce the demanded computation time [5]. A very useful side effect of the simplifications employed is the fact that they open the possibility of longer time periods to be chosen for the sampling of the dynamic simulation. This offers a way of observing the dynamic behavior of large molecular systems more completely.

Before discussing the various possibilities in detail it must be mentioned again that each modification and reduction of the number of degrees of freedom can cause a lack of accuracy and it has to be checked carefully whether or not a respective simplification can be tolerated.

One basic and very common simplified procedure is the use of *united atom potential energy functions*. The underlying theory of this methodology has been described earlier. Most of the force fields for protein modeling, such as AMBER [1, 2] and GROMOS [3] are based on these algorithms. Omission of the non-polar hydrogens in a united-atom force field does significantly reduce the number of particles in a large biomolecule. A further possibility to reduce the demand for computer time is provided by application of the SHAKE [8] algorithm. In the SHAKE procedure additional forces are assigned to the atoms, aiming to keep bond lengths fixed at equilibrium values. This is very useful for several reasons. Above all, bond stretching energy terms must not be calculated for the frozen bonds. The magnitude of the integration step depends on the fastest occurring vibrations in a molecule. This is usually the high frequency vibration of the C–H bond stretching. This period is of the order of 10^{-14} seconds; therefore the integration step should be chosen to be 10^{-15} seconds (1 femtosecond). Applying the SHAKE algorithm to this type of C–H bond allows a larger integration step with the effect of reducing the necessary computational expense, and thereby offering the chance of simulating the system over a longer time period. The definition of cut-off radii, leading to a neglect of non-bonded interactions beyond the defined distance, also yields the same effect.

In addition, the application of a well-balanced computational protocol may save computer time. In this respect several parts of a protein can be kept rigid and the molecular dynamics simulations then carried out only for flexible parts such as loops or side chains, while well-defined secondary structures like α-helices or β-strands in the core of the protein are not taken into account. The availability of NMR data can also be a reason to fix atoms, side chains or parts of the protein at their initial coordinates in order to impede their movement away from the experimentally derived positions. Again, a warning must be given; restraining parts of flexible molecules leads to a reduction in the number of degrees of freedom. Without any doubt a more comprehensive exploration of the conformational space, and hence better results, are achieved when no positional restraints are applied on parts of the protein structure.

All mentioned methods enhance the efficiency of the molecular dynamics simulations; nevertheless, for some problems the feasible time scale is still too short. If for example the binding of a ligand to an enzyme or receptor protein—as well as the thereby triggered conformational change—is to be studied, the time required for this process can be in the order of picoseconds or even nanoseconds [9]. The same time-scale would be indispensable for a simulation of protein folding. Both types of problems are still out of reach.

Several modifications of high-temperature molecular dynamics simulations have been successfully applied in conformational analysis of peptides and in the refinement of protein models. In this respect two important methods, the high-temperature annealed molecular dynamics simulation [10] and simulated annealing [11] have been discussed in section 2.3.3. They are valuable tools and widely employed also for investigating peptides and proteins [12–16].

A sensitive point in all molecular dynamics protocols is the choice of the suitable simulation temperature. Usually the simulation will be performed in the range between 300 K and 400 K. On one hand it must be sufficiently high to prevent the system from getting stuck in one particular region of the conformational space, but on the other hand should not be too high, as this could result in distorted high-energy conformations, even after minimization [16]. Another commonly observed problem in the application of high-temperature molecular dynamics simulations of proteins and peptides is the appearance of *trans–cis* interconversions of peptide bonds. These artefacts can be avoided by using lower temperatures at the expense of conformational search efficiency, or by introducing torsional restraints onto the peptide bonds.

4.4.4 Treatment of Solvated Systems

The conformational flexibility of a protein especially at the surface and in loop regions is strongly affected by the surrounding environment. The non-existence of neighboring atoms at the surface of the protein leading in effect to in vacuo conditions for these regions of the protein and the problems associated therewith have already been mentioned when discussing the minimization process. Of course the accuracy of molecular dynamics simulations increases by including explicit solvent. Unfortunately, this is still an unresolved problem. One possibility to mimic the effect of solvent and to account for the boundary phenomena is the use of distance-dependent dielectric constants.

To enwrap the molecule with a sphere of solvent molecules can improve the accuracy of the molecular dynamics simulations because by doing this at least part of the effect of solvation are imitated. At this point it is important to note that there is a decisive difference between simple solvent water and structural water. Structural water is very important for the functionality of the protein and can mainly influence its conformation, even in the core. Therefore structural water must always be included explicitly in the calculations.

The next level of improvement is the embedding of the protein in a complete solvent box containing thousands of water molecules in order to simulate a natural solvent environment. This is not always possible because the required computational effort is immense but has been applied successfully in a recent molecular dynamics study of a complete enzyme [17]. A comprehensive review on molecular dynamics simulations of proteins in different environments is given in [18].

In the majority of cases the use of realistic water models with thousands of molecules is too time-consuming. For this reason specific methods using a simplified representation of solvent molecules have been developed [18]. Solvent molecules for example can be substituted by neutral spherical atoms. This type of procedure significantly reduces the computational effort. A detailed discussion of all procedures used in this context is beyond the scope of this book; nevertheless, it is important to consider that inclusion of the solvent environment at any level of complexity into the calculations is an important means for improving the accuracy and the reliability of molecular dynamics simulations, especially for large biomolecular systems.

4.4.5 Ligand-Binding Site Complexes

Generated protein models are often used for studying ligand–protein interactions. Small molecules which are mostly new drugs of pharmaceutical interest can be placed into the active site of the protein. As the natural binding process is not static, molecular dynamics simulations are necessary to simulate the dynamical properties of the ligand–protein complex. Valuable information like hydrogen bonding pattern, rms deviations and positional fluctuations can be deduced from the simulation to discriminate between binding and non-binding ligands.

Several prerequisites must be fulfilled for a meaningful molecular dynamics simulation of ligand–protein complexes. The initial coordinates both of the ligand and the protein must represent an energetically reasonable conformation. The simulated system must include all regions of interest and must be large enough to sample all forces contributing to the total energy of the system correctly. Truncated active site complexes can only be studied if all possible ligand–protein interactions can be reflected during the molecular dynamics simulation. Last, but not least, the simulation time must be sufficiently long in order to generate a representative ensemble of data.

In spite of the known limitations, molecular dynamics simulations have also become a powerful tool for investigating dynamical processes of biopolymers such as peptides, proteins, enzymes, receptors and membranes. The combination of experimental results like NMR measurements or crystal data with theoretical methods can provide a route for gaining a detailed 3D atomic picture of the molecular system and to study hitherto experimentally inaccessible processes in proteins.

References

[1] Weiner, S. J., Kollman, P. A., Case, D. A., Singh, U. C., Ghio, C., Alagona, G., Profeta, S. Jr., and Weiner, P. *J. Am. Chem. Soc.* **106**, 765–784 (1984).

[2] Weiner, S. J., Kollman, P. A., Nguyen, D. T., and Case, D. A. *J. Comput. Chem.* **7**, 230–252 (1986).

[3] van Gunsteren, W. F., and Berendsen, H. J. C. Molecular dynamics simulations: techniques and applications to proteins. In: *Molecular Dynamics and Protein Structure*. Hermans, J. (Ed.). Polycrystal Books Service: Western Springs; 5–14 (1985).

[4] Brooks, C. L., III, Montgomery, Pettitt, B., and Karplus, M. *J. Chem. Phys.* **83**, 5897–5908 (1985).

[5] van Gunsteren, W. F. *Adv. Biomol. Simul.* **239**, 131–146 (1992).

[6] Dauber-Osguthorpe, P., Roberts, V. A., Osguthorpe, D. J., Wolff, J., Genest, M., and Hagler, A. T. *Proteins: Structure, Function and Genetics* **4**, 31–47 (1988).

[7] Brooks, B. R., Bruccoleri, R. E., Olafson, B. D., States, D. J., Swaminathan, S., and Karplus, M. *J. Comput. Chem.* **4**, 187 (1983).

[8] Ryckaert, J. P., Ciccotti, G., and Berendsen, H. J. C. *J. Comput. Phys.* **23**, 327 (1977).

[9] Lybrand, T. P. Computer Simulation of Biomolecular Systems Using Molecular Dynamics and Free Energy Perturbation Methods. In: *Reviews in Computational Chemistry*, Vol. 1. Lipkowitz, K. B., and Boyd, D. B. (Eds.). VCH: New York; 295–320 (1990).

[10] Auffinger, P., and Wipff, G. *J. Comput. Chem.* **11**, 19–31 (1990).

[11] Kirkpatrick, S., Gelatt, C. D., and Vecchi, M. P. *Science* **220**, 671–680 (1983).

[12] Salvino, J. M., Seoane, P. R., and Dolle, R. E. *J. Comput. Chem.* **14**, 438–444 (1993).

[13] Wilson, S. R., and Cui, W. *Biopolymers* **29**, 225–235 (1990).

[14] Mackey, D. H. J., Cross, A. J., and Hagler, A. T. The Role of Energy Minimization in Simulation Strategies of Biomolecular Systems. In: *Prediction of Protein Structure and the Principles of Protein Conformation*. Fasman, G. (Ed.). Plenum Press: New York; 317–358 (1989).

[15] Kerr, I. D., Sankararamakrishnan, R., Smart, O. S., and Sansom, M. S. P. *Biophys. J.* **67**, 1501–1515 (1994).

[16] Bruccoleri, R. E., and Karplus, M. *Biopolymers* **29**, 1847–1862 (1990).

[17] Vijayakumar, S., Ravishanker, G., Pratt, R. F., and Beveridge, D. L. *J. Am. Chem. Soc.* **117**, 1722–1730 (1995).

[18] van Gunsteren, W. F., and Karplus, M. *Biochemistry* **21**, 2259–2274 (1982).

4.5 Validation of Protein Models

Once a protein model has been built using knowledge-based methods and subsequently optimized by molecular mechanics or molecular dynamics, it is important to assess its quality and reliability. The question arises how protein models can be tested for correctness and accuracy. This is a very difficult business, because the quality of a homology-based protein model depends on a huge number of properties on different levels of structural organization. This is summarized in Fig. 1.

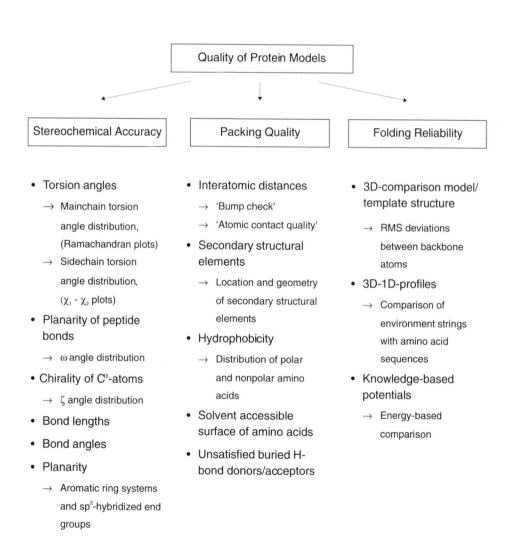

Figure 1. Quality questionnaire for protein models.

4.5.1 Stereochemical Accuracy

The quality of the 3D structure of a protein model depends strongly on the accuracy of the used template structure, i.e. the quality of the crystal structure [1]. Of course, the modeled protein cannot show higher accuracy than the crystal structure which has been used as a template. Protein structures derived from X-ray diffraction can contain errors, both experimental and in the interpretation of the results [1–3]. The general measures for the quality of crystal structures are the resolution and the R-factor. The better the resolution of the protein crystal the greater the number of independent experimental observations derived from the diffraction data and hence the greater the accuracy of the protein structure [4]. The resolution of protein structures contained in the Brookhaven database is usually found to be in the range of 1–4 Å. The R-factor is a measure for the agreement between the derived 3D structure of a protein crystal (the 3D structure which fits the electron density map best) and the "real" crystal structure. The R-factor can be determined by comparing the experimentally obtained amplitudes of the X-ray reflections and the amplitudes calculated from the protein structure which shows the best fit to the electron density map (for a detailed discussion about the accuracy of protein X-ray crystallography the reader is referrred to the literature [5]). The better the agreement between observed and calculated amplitudes (resulting in a low R-factor), the better the agreement between the derived and the real crystal structure. The R-factor can be artificially reduced in a number of ways and therefore sometimes might be misleading [2]. It is commonly accepted to consider structures with a resolution of 2.0 Å or better to be reliable. If in addition the R-factor is below 20% it can be safely assumed that the protein structure is essentially correct.

To verify the stereochemical quality of a model-built structure, the accuracy of parameters such as bond lengths, bond angles, torsion angles and correctness of the amino acid chirality, must be proved. It has been observed in 3D structures of proteins that mainly the bond lengths and angles cluster around the "ideal values". Thus, the mean values detected in crystal structures can be regarded as good indicators of the stereochemical quality and must be compared with the actual values in the generated protein model (see Table 1) [6] in order to discover stereochemical irregularities which would disclose a bad structure.

Since a manual inspection of all stereochemical parameters of a protein will be tedious and time-consuming, programs have been developed which automatically check all stereochemical properties. Examples are PROCHECK [7] or WHATCHECK [8], both of which are available free of charge (see http://www.biotech.embl-heidelberg.de).

One important indicator of stereochemical quality is the distribution of the main chain torsion angles ϕ and ψ. The distribution of all ϕ and ψ torsion angles in a protein can be examined in a Ramachandran plot. As described in section 4.2.1, the favored and unfavored regions of the classical Ramachandran plot have been determined by studying the conformational behavior of isolated dipeptides. Very conveniently the ϕ–ψ torsion angles observed in hundreds of well-refined protein structures generally lie within the same regions as determined for the isolated dipeptides. It is one of the remarkable properties of repetitive secondary structures in proteins that the observed ϕ,ψ-values are very close to the optimal dipeptide conformations, as calculated by Ramachandran. Also the ϕ and ψ torsion angles of non-repetitive structures, like loops or turns, are found within the favored regions of the Ramachandran plot, but are more widely distributed over these areas.

Table 1. Stereochemical parameters derived from high-resolution protein structures after Morris et al. [6]

Stereochemical parameters	Mean Value	Standard deviation
ϕ–ψ in most favored regions of Ramachandran plots	> 90%	–
χ_1 torsion angle gauche minus	64.1°	15.7°
trans	183.6°	16.8°
gauche plus	–66.7°	15.0°
χ_2 torsion angle	177.4°	18.5°
Proline ϕ torsion angle	–65.4°	11.2°
α-Helix ϕ torsion angle	–65.3°	11.9°
α-Helix ψ torsion angle	–39.4°	11.3°
Disulfide bond separation	2.0 Å	0.1 Å
ω Torsion angle	180.0°	5.8°
C^α tetrahedral distortion: z torsion angle (virtual torsion angle C^α–N–C–C^β)	33.9°	3.5°

As an example, the Ramachandran plot of a protein crystal structure (cephalosporinase of *enterobacter cloacae*) is shown in Fig. 2. The torsion angles of all residues, except those for proline residues and those at the chain termini, are presented. Glycine residues are separately identified by triangles (as those are not restricted to any particular region of the plot). The shading represents the different major regions of the plot: the darker the region the more favored is the corresponding ϕ,ψ combination. The white region is the disallowed region for normal amino acids and any residue found in this region must be carefully inspected. Usually amino acids lying in less-favored regions are especially labelled (in Fig. 2 shown in red) with residue name and residue number for easy indentification and inspection.

Unfavorable stereochemistry, becoming visible by disallowed ϕ,ψ torsion angles, seems to occur in natural proteins exclusively if the special geometry is required for function or stability, for example when residues in the core of the protein are involved in hydrogen bonds or salt-bridges. Residues which are allowed to lie outside the major regions of the Ramachandran plot are proline and glycine. Because glycine and proline have—due to their different stereochemistry—other favored and unfavored regions, it is more convenient to mark these amino acid types particularly or to exclude them from the normal Ramachandran plot. Therefore, separate Ramachandran plots for all glycines, all prolines, and all other amino acids are very often created. The percentage of residues lying in the favored regions of a Ramachandran plot is one of the best guides to check stereochemical quality of a protein model. Ideally, one would hope to have more than 90% of the residues in the allowed regions [7].

The same check as described above for main chain torsion angles can be applied in case of the side chain torsion angles χ_i. The χ_1 torsion angles, observed in well-refined structures of proteins [6], are generally close to one of the three possible staggered conformations, the most

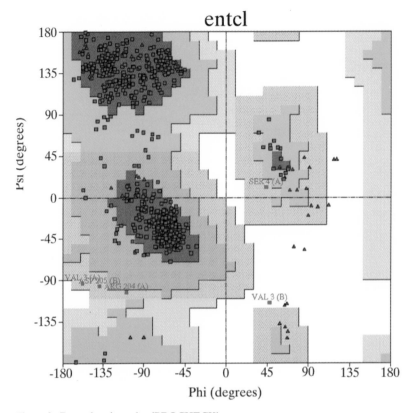

Figure 2. Ramachandran plot (PROCHECK).

favored conformation being the one where the bulkiest groups are most remote (see Table 1: gauche plus, trans, and gauche minus torsion angles for χ_1). For the χ_2 torsion angles a preference for the trans conformation has been found. A similar distribution for the side chain torsion angles in protein crystal structures has been detected by Ponder and Richards [9]. The distribution of the side chain torsion angles of all amino acid types in protein models can be inspected in more detail in graphs, where usually side chain torsion angles χ_1 are plotted versus χ_2. Examples for this kind of graph are shown in Fig. 3 for a cephalosporinase of *Enterobacter cloacae*. Every single plot shows the $\chi_1-\chi_2$ angle distribution for a particular amino acid type. The green shading on each plot indicates the favorable regions which have been determined from a data set of well-resolved protein crystal structures [7]. Black marks indicate the corresponding values found in the cephalosporinase; the red marks denote outliers.

Some of the stereochemical parameters of protein structure have been found to be constant in all known proteins. Of course, these properties are a very sensitive measure for the quality of protein models and must be carefully checked for consistency.

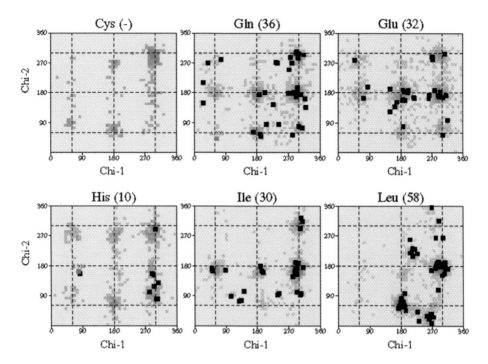

Figure 3. $\chi_1-\chi_2$ Plot for different amino acids (PROCHECK).

This list contains:

1. *The peptide bond planarity:* this property is usually measured by calculating the mean value and the standard deviation of all ω angles in the investigated protein. The smaller the standard deviation, the tighter the clustering around the normal value of 180°, which represents the planar trans configuration (see also Table 1 for the distribution of ω angles in crystal structures). All *cis* peptide bonds are also separately listed and must be inspected. *Cis* peptide bonds occur in proteins at about 5% of the bonds that precede proline residues. Regarding all peptide bonds, which do not involve proline residues, the *cis* configuration is observed less than 0.05% [10, 11].
2. *The chirality of the C^α-atoms:* One of the general principles of protein structure is the preference for one handedness over the other (e.g. the preference for the right-handed conformation of an α-helix). The basis for this is the presence of an asymmetric center at the C^α-atom which in all naturally occurring amino acids is L-configurated. A protein model must therefore be examined for correct chiralities. A parameter which provides a measure for the correctness of chirality is the ζ torsion angle. This is a virtual torsion angle which is not defined by any actual bond in a protein. Rather this torsion angle is determined by the C^α–N–C'–C^β atoms of each amino acid residue. The numerical values of the ζ torsion angle should reside between 23° and 45°. If the value is negative, this fact signifies the appearance of an incorrect D-amino acid [7].

3. *Main chain bond lengths and angles:* the distribution of each of the different main chain bond lengths and angles in a protein is compared with the distribution observed in well-resolved crystal structures. Usually deviations more than 0.05 Å for bond lengths and 10° for bond angles are regarded as distorted geometries which have to be inspected in detail [3].

Aromatic ring systems (Phe, Tyr, Trp, His) and sp^2-hybridized end groups (Arg, Asn, Asp, Glu, Gln) must be checked for planarity. The deviation of these parameters, i.e. the distorted geometry, is often the result of bad interatomic contacts. Removing the steric constraints and subsequently optimizing the model in most cases yields a relaxed structure with ideal geometrical parameters.

4.5.2 Packing Quality

Specific packing interactions within the interior are assumed to play an important role for the structural specifity of proteins [12–14]. It has been observed that globular proteins are tightly packed with packing densities comparable with those found in crystals of small organic molecules [12]. The interior of globular proteins contains side chains that fit together with striking complementarity, like pieces of a 3D jigsaw puzzle. The high packing densities observed in proteins are the consequence of the fact, that segments of secondary structure are packed together closely; helix against helix, helix against strands of a β-sheet, and strands against strands of different β-sheets [14–17]. The interior packing of globular proteins is a major contribution to the stability of the overall conformation. Therefore, the packing quality of a protein model can be used to estimate its reliability. It can be judged using a variety of methods, which will be described in detail in this section.

In a first step, it must be verified that the generated and refined protein model includes no bad van der Waals contacts. Therefore, all interatomic distances must be examined for residing in ranges which have been observed in well-refined crystal structures. Several procedures exist for this distance check. In the simplest, all interatomic distances are measured and those with distances below a determined threshold are defined as bad contacts which have to be inspected in detail (for example, 2.6 Å is used as threshold in the PROCHECK program [7]). A more accurate judgement of interatomic distances is performed by programs like WHATCHECK [8]. For all well-refined protein crystal structures stored in the Brookhaven database all interatomic distances shorter than the sum of their van der Waals radii +1.0 Å are determined and stored. The distance that subdivides the collected values such that 5% of all observed distances are shorter and 95% are longer than this measure, is defined as "short normal distance". As there are 163 different atom types in the naturally occuring amino acids 163 × 163 "short normal distances" are defined. All distances occuring in the protein model which are more than 0.25 Å shorter than the short normal distances are reported by the program.

The next step involves the examination of the secondary structural elements of the protein model. As we have already mentioned in section 4.3.2, the secondary structural elements are the most conserved regions in highly homologous proteins. Thus, it must be proven whether the secondary structural elements observed in the template protein also can be detected in the protein model, i.e. whether the secondary structure has been maintained during the building and optimization process. Programs which can be applied for this purpose are the DSSP [18] or the STRIDE program [19] (see section 4.3.2). These programs allow a more sophisticated assignment of secondary structure than the manual inspection of α-helices and β-sheets.

A variety of methods exist, which use the huge amount of information derived from protein crystal structures to estimate the packing quality of model built structures [20–23]. From the assumption that atom–atom interactions are the primary determinant of protein conformation, Vriend and Sander have developed a program that checks the packing quality of a protein model by calculating a so-called "contact quality index" [20]. This index is a measure of the agreement between the distributions of atoms around an amino acid side chain in the protein model and equivalent distributions observed in well-resolved protein structures. For that reason a database has been generated which contains a contact probability distribution for all amino acid side chains. This magnitude describes the probability for a certain atom type to occur in a particular region around the side chain. These probability values are used to check the contact quality in the protein model. The better the agreement between the distributions in the model and in the crystal structures the higher the contact quality index, and the more favorable the residue packing.

The distribution of polar and non-polar residues between the interior and the surface of proteins has been found to be a general principle of the architecture of globular proteins. At a simple level, a globular protein can be considered to consist of a hydrophobic interior surrounded by a hydrophilic external surface which interacts with the solvent molecules. These building principles have been identified in most 3D structures of globular proteins and can be summarized as follows:

1. The interior of globular proteins is densely packed without large empty space and is generally hydrophobic. Non-polar side chains predominate in the protein interior; Val, Leu, Ile, Phe, Ala, and Gly residues comprise 63% of the interior amino acids [10]. Ionized pairs of acidic and basic groups hardly occur in the interior, even though such pairs might be expected to have no net charge due to the formation of salt-bridges.

2. Charged and polar groups are located on the surface of globular proteins accessible to the solvent. On average, Asp, Glu, Lys, and Arg residues comprise 27% of the protein surface and only 4% of the interior residues [10]. (Integral membrane proteins differ from globular proteins primarily in having extremely non-polar surfaces which are in contact with the hydrophobic membrane core.)

These features make a major contribution to the stability of folded proteins [14, 24, 25]. The underlying principle for this distribution is the hydrophobic effect, i.e. the removal of hydrophobic residues from contact with water. It has been observed that the free energies, associated with the transfer from water to organic solvent, of polar, neutral and non-polar residues are correlated with the extent to which they occur in the interior and exterior of proteins [26]. Therefore, the distribution of hydrophobic and hydrophilic residues in proteins, can be used to estimate the reliability of protein models [26–29]. Several programs have been developed which use this feature as a measure of the packing quality [8, 28, 29] of a protein model.

It has been also observed that the hydrophobicity of an amino acid (defined as free energy of transfer from water to organic solvent) is related linearly to its surface area, i.e. the more hydrophobic the residue, the more completely buried it will be [30]. The buried surface area of a particular amino acid is herein defined as the difference between the solvent-accessible surface of the residue in an extented polypeptide chain (usually defined as the "standard state" in the tri-peptide Gly-XXX-Gly) and the solvent-accessible surface of the residue in the folded protein. It has been demonstrated that the buried surface area, i.e. the area which is lost when a residue is transferred from the defined "standard state" to a folded protein, is proportional to its hydrophobicity.

Additionally, the total surface buried within globular proteins has been found to correlate with their molecular weights, i.e. upon folding, globular proteins bury a constant fraction of their available surface [26]. Several programs have been developed which use the general properties of amino acid surfaces in order to provide an estimation of the packing quality of globular proteins [8, 28, 30]. For a detailed review of the topic of molecular surfaces and their contributions to protein stability the reader is referred to the literature [14, 31].

Although the residues that form the protein interior are usually non-polar or neutral, there are rare cases of buried polar residues. It has been observed in many investigations of protein crystal structures that virtually all polar groups in the protein interior are paired in hydrogen bonds. Many of these polar groups form hydrogen bonds within their own secondary structure (i.e. α-helices and β-sheets). Others are involved in binding co-factors, metal ions or are located in the active site of proteins. Buried ionizable groups, which rarely occur inside globular proteins, are usually always involved in salt-bridges. Sometimes the positive and negative charges are bridged by water molecules. Due to this observation, it is necessary to check the protein model whether all polar buried residues are paired in hydrogen bonds and whether all charged residues are involved in salt-bridges. Salt-bridges and hydrogen bonds are usually identified on the basis of their interatomic distances [32].

4.5.3 Folding Reliability

Proteins with homologous amino acid sequences generally have similar folds. Therefore the overall 3D structure of the protein model and its template should be similar. Especially in the structurally conserved regions, the homologous proteins must possess the same conformation. In cases where the originally constructed protein model contains large regions of steric strain (due to the incorrect architecture), the protein may undergo correspondingly large movements in its 3D structure during the refinement process. The resulting protein conformation is not reliable, because it shows only little agreement with the 3D structure of the template protein. When checking protein conformations, one normally measures the similarity in 3D structure by the rms deviations of the C^{α}-atomic or the backbone coordinates after optimal rigid body superposition of the two structures (for details, see [33]). A very large rms deviation means the two structures are dissimilar; a value of zero means that they are identical in conformation. Homologous proteins generally show low rms deviations for their C^{α}-atoms, but no general value exists which can be used as an indicator whether two protein structures are similar or dissimilar. Chothia and Lesk [34] have performed an investigation on structural similarity of homologous proteins. The overall extent of the structural divergence of two homologous proteins was measured by optimally superposing the common conserved regions (the so-called common core) and calculating the rms difference in the positions of their backbone atoms. For a test set of 32 homologous pairs of proteins they have found rms differences for the common cores which vary between 0.62 and 2.31 Å (see Fig. 4).

When the overall structural similarity of the protein model and the template protein have been evaluated, the question arises whether the generated conformation for an unknown protein is the correct native fold. How can one prove whether the constructed model is correct in its overall conformation? In the search for criteria that discriminate between the correct conformation and incorrectly folded models Novotny et al. have performed an interesting investigation [35]. They have studied two structurally dissimilar but identically large proteins, hemerythrin (1HMQ) and the variable domain of mouse immunoglobulin κ-chain (1MCP-L). The two proteins have been modified by placing the amino acid sequence of one protein on to the backbone structure of the other and vice versa, in order to obtain incorrect models. The model structures were optimized to remove steric overlaps of side chains. After minimization, the total energies of native folds and incorrect protein models were approximately the same. The authors concluded that the energies obtained from standard force field calculations cannot be used to distinguish between correct and incorrect protein conformations. On the other hand, the investigation has shown that the packing criteria of the incorrect models were different from those normally found in native proteins. The incorrect structures clearly violated the general principles of close packing, hydrogen bonding, minimum exposed non-polar surface area, and solvent accessibility of charged groups. Examination of the interior

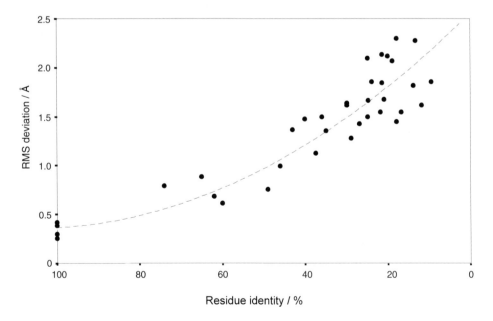

Figure 4. rms/sequence identity plot.

showed that the packing of side chains at the secondary structure interfaces also differed from the characteristics observed in natural proteins (e.g. side chain ridges and grooves spirally wound on α-helices, predominantly flat surfaces of β-sheets). This analysis has clearly shown that the validity of model-built structures can only be assessed by a careful inspection of the structural features of a protein model.

For that reason several methods have been developed which try to distinguish between correct and incorrect folded protein structures [36–44]. One of these approaches is the 3D-Profiles method [36–39], which is based on the general principle that the 3D structure of a protein must be compatible with its own amino acid sequence. It measures this compatibility by reducing the 3D structure of the protein model to a simplified one-dimensional (1D) representation, the so-called *environment string*. The environment string has the same length as the corresponding amino acid sequence. This 1D string then can be compared with the respective amino acid sequence, which is also a 1D parameter.

In a first step the 3D structure of the protein model must be converted into a 1D parameter. For that reason the program determines several features of the environment of each residue: the area of the side chain that is buried in the protein; the fraction of the side chain area that is exposed to polar regions; and the secondary structure to which the particular amino acid belongs. Based on these characteristics each residue position is categorized into an environment class. A total of 18 distinct environment classes are implemented in the program [38]. In this manner the 3D structure is translated into a 1D string which represents the environment class of each residues in the protein model.

Although the environment string is 1D, it cannot be aligned with an amino acid sequence without some measure of compatibility for each of the distinct environment classes with each of the 20 naturally occurring amino acids. For that reason the program includes a compatibility scoring matrix (comparable with the scoring matrices described in section 4.3.1), which has been derived from sets of known protein structures [39]. Applying this compatibility matrix, the environment string and the amino acid sequence are aligned and a so-called 3D–1D score is obtained for the particular alignment. For obvious reasons it is more convenient to calculate local 3D–1D scores for small and medium-sized regions of about 5–30 residues length, than a global score for the complete alignment. The local scores are then plotted against residue positions to reveal local regions of relatively high or low compatibility between the 3D structure and the amino acid sequence [38]. Regions showing unusually low scores are likely to be regions where the protein conformation is incorrect, or where structural refinement is necessary.

The folding reliability can be also tested using knowledge-based force field methods [42–44]. These methods are based on the compilation of potentials of mean forces from a database of known 3D protein structures. The basic idea of these approaches is that atom–atom interactions in proteins are the primary determinant of proper protein folding.

A program named PROSA-II has been developed, which uses the mean force potentials to calculate the total energy of amino acid sequences in a number of different folds [42]. The calculated total energy of a particular protein conformation is a qualitative criterion for the confidence or quality of a predicted protein model. This is in contrast to investigations where the total energies, derived from standard molecular mechanics force fields, have been used to estimate the reliability of different protein conformations [35]. To test the predictivity of PROSA-II different native and incorrectly modified protein conformations have been used as a test set. It has been shown that for a very large number of proteins the derived total energy of the correctly folded protein is much lower than for any alternative (incorrect) protein conformation. Therefore, the program can be successfully applied to recognize erroneous protein folds or to detect faulty parts of structures in protein models.

References

[1] Bränden, C. J., and Jones, T. A. *Nature* **343**, 687–689 (1990).

[2] Jones, T. A., Zou, J. Y., and Cowan, S. W. *Acta. Cryst.* **A47**, 110–119 (1991).

[3] Engh, R. A., and Huber, R. *Acta. Cryst.* **A47**, 392–400 (1991).

[4] Hubbard, T. J. P., and Blundell, T. L. *Protein Eng.* **1**, 159–171 (1987).

[5] Drenth, J. *Principles of Protein X-ray Crystallography*. Springer Verlag: New York 1994.

[6] Morris, A. L., MacArthur, M. W., Hutchinson E. G., and Thornton, J. M. *Proteins* **12**, 345–364 (1992).

[7] Laskowski, R. A., MacArthur, M. W., Moss, D. S., and Thornton, J. M. *J. Appl. Cryst.* **26**, 283–291 (1993).

[8] WHAT IF, Vriend, G., European Molecular Biology Laboratory (EMBL), Heidelberg, Germany.

[9] Ponder, J., and Richards, F. M. *J. Mol. Biol.* **193**, 775–791 (1987).

[10] Creighton, T. E. *Proteins: Structures and Molecular Properties*. 2nd Ed. W. H. Freeman and Company: New York 1993.

[11] Stewart, D. E., et al. *J. Mol Biol.* **214**, 253–260 (1990).

[12] Richards, F. M. *J. Mol. Biol.* **82**, 1–14 (1974).

[13] Richards, F. M. *Annu. Rev. Biophys. Bioeng.* **6**, 151–176 (1977).

[14] Chothia, C. *Annu. Rev. Biochem.* **53**, 537–572 (1984).

[15] Zehfus, M. H., and Rose, G. D. *Biochemistry* **25**, 5759–5765 (1986).

[16] Janin, J., and Chothia, C. *J. Mol. Biol.* **143**, 95–128 (1980).

[17] Leszczynski, J. F., and Rose, G. D. *Science* **234**, 849–855 (1986).

[18] Kabsch, W., and Sander, C. *Biopolymers* **22**, 2577–2637 (1983).

[19] Frishman, D., and Argos, P. *Proteins Struct. Func. Gen.* **23**, 556–579 (1995).

[20] Vriend, G., and Sander, C. *J. Appl. Cryst.* **26**, 47–60 (1993).

[21] Gregoret, L. M., and Cohen, F. E. *J. Mol. Biol.* **211**, 959–974 (1990).

[22] Singh, J., and Thornton, J. M. *J. Mol. Biol.* **211**, 595-615 (1990).

[23] Privalov, P. L., and Gill, S. J. *Adv. Protein Chem.* **39**, 191–234 (1988).

[24] Chothia, C. *J. Mol. Biol.* **105**, 1–12 (1976).

[25] Wolfenden, R., Anderson, L., Cullis, P. M., and Southgate, C. B. *Biochemistry* **20**, 849 (1983).

[26] Miller, S., Janin, J., Lesk, A. M., and Chothia, C. *J. Mol. Biol.* **196**, 641–656 (1987).

[27] Lee, B., and Richards, F. M. *J. Mol. Biol.* **55**, 379–400 (1971).

[28] Eisenberg, D., and McLachlan, A. D. *Nature* **319**, 199–203 (1986).

[29] Baumann, G., Frömmel, C., and Sander, C. *Protein Eng.* **2**, 329–334 (1989).

[30] Rose, G. D., Geselowitz, A. R., Lesser, G. L., Lee, R. H., and Zehfus, H. *Science* **229**, 834–838 (1985).

[31] Rose, G. D., and Dworkin, J. E. The Hydrophobicity Profile. In: *Prediction of Protein Structure and Function and the Principles of Protein Conformation*. Fasman, G. D. (Ed.). Plenum Press: New York; 625–634 (1989).

[32] Rashin, A., and Honig, B. *J. Mol. Biol.* **174**, 515–521 (1984).

[33] Rao, S. T., and Rosman, M. G. *J. Mol. Biol.* **76**, 214–228 (1973).

[34] Chothia, C., and Lesk, A. M. *EMBO J.* **5**, 823–826 (1986).

[35] Novotny J., Bruccoleri, R., and Karplus, M. *J. Mol. Biol.* **177**, 787-818 (1984).

[36] Gribskov, M., McLachlan, A. D., and Eisenberg, D. *Proc. Natl. Acad. Sci. U.S.A.* **84**, 4355–4358 (1987).

[37] Bowie, J. U., Lüthy, R., and Eisenberg, D. *Science* **253**, 217–221 (1990).

[38] PROFILES-3D, User Guide, Biosym Technologies, San Diego, California, USA

[39] Lüthy, R., McLachlan, A. D., and Eisenberg, D. *Proteins* **10**, 229–239 (1991).

[40] Novotny, J., Rashin, J. J., and Bruccoleri, R. E. *Proteins Struct. Funct. Genet.* **4**, 19–25 (1988).

[41] Hendlich, M., Lackner, P., Weitckus, S., Floeckner, H, Froschauer, R., Gottsbacher, K, Casari, G., and Sippl, M. J. *J. Mol. Biol.* **216**, 167–180 (1990).

[42] Sippl, M. J. *Proteins* **17**, 355–362 (1993).

[43] Casari, G., and Sippl, M. J. *J. Mol. Biol.* **224**, 725–732 (1992).

[44] MacArthur, M. W., Laskowski, R. A., and Thornton, J. M. *Curr. Opin. Struct. Biol.* **4**, 731–737 (1994).

4.6 Properties of Proteins

4.6.1 Electrostatic Potential

As we have already mentioned electrostatic interactions are among the most important factors in defining the conformation of a molecule in aqueous solution and in determining the energetics of interaction between two approaching molecules. The protein itself, the solvent, cofactors and prosthetic groups are nearly always charged or dipolar, and so the range of effects which are dependent in one way or another from electrostatics is broad [1–4]. Contrary to dispersion forces the electrostatic interactions are effective over relatively large distances. Due to their strong influence on structure and function of macromolecules in aqueous solution, it is absolutely necessary to consider explicitly the electrostatic term in any theoretical study on proteins [1]. For this purpose theoretical models are needed, which are able to describe correctly the electrostatic effects in proteins.

The interaction between any two charges is described by Coulomb's law (see section 2.2.1) which, in its simplest form is only valid for two point charges in vacuum. If the charges are immersed in any other matter, then particles of the surrounding matter are polarized by the presence of the charges, and the induced dipoles of the particles interact with the original point charges. Thus, the total resolved force on each of the point charges is altered, and the electrostatic interaction is decreased by the influence of the dielectric medium.

In classical electrostatic approaches, the materials are considered to be homogeneous dielectric media, which can be polarized by charges and dipoles. A dielectric constant is used as a macroscopic measure of the polarizability of a medium rather than explicitly accounting for the polarization of each atom. The portrayed procedure is called a *continuum model*.

It must be borne in mind that this view is simplistic and that the concept of dielectric constant—which constitutes a genuine macroscopic property—is valid only for homogeneous media. Less homogeneous environments must be treated explicitly. Special problems arise at the boundaries between regions of very different dielectric properties [5]. The surface of a protein represents such a case, because it divides the molecule into two regions which differ dramatically in composition. The molecular interior possesses a very low dielectric constant and includes a particular number of charges (most of them near the surface). Outside the protein there is a polar aqueous medium, which normally contains a distinct quantity of ions. For two point charges separated by a specific distance in a macromolecule in aqueous solution, the electrostatic interaction energy depends on the shape of the macromolecule and the exact positions of the charges (for a detailed description of this topic, see [5, 7, 8]. When using Coulomb's law for the calculation of electrostatic interactions, this fact will not be taken into consideration.

The multiple interactions occurring among the point charges and dipoles of the protein and the solvent are mutually dependent and turn the simple relationship of Coulomb's law into a very complex state. The electrostatic interaction among molecules in a homogeneous environment can be averaged and expressed, as we have seen above, with the help of a simple dielectric constant. This concept is not valid for the inhomogeneous environment of proteins. Their electrostatic properties involve interactions among the multiple charges and dipoles of the proteins, and between these and the surrounding solvent and any ions that it contains. In this situation interactions between particular charges and dipoles must be calculated individually. This is impractical with the many atoms of the protein and solvent.

The major problem studying electrostatic effects in proteins is, as we have seen, the treatment of polarization effects. In many electrostatic problems, real materials are treated as simple continua, and the effects of the underlying microscopic structure of the material is only incorporated into the macroscopic dielectric constant. At the microscopic level, the shielding of the charges arises from the polarizability of the individual atoms. Thus, an approach which discards the use of a dielectric constant and considers the individual atoms of the system and their mutual polarizabilities would be the best way to solve the problem. Of course, the exact quantum mechanical treatment would be a suitable solution, but this at present—due to limitations of computer power—is not practicable for systems of the size of proteins. Therefore, empirical approaches are generally employed for the exact calculation of electrostatic interactions within proteins [6–12].

Most of these approaches make use of the point charge approximation, i.e. the charge distribution of a protein is described by locating point charges at the atom centers. Several methods have been developed to obtain corresponding partial charges [13–15]. The procedures used are comparable with those described for the small molecules (see section 2.4.1.1). Because the complete protein is too large for a quantum mechanical charge calculation, the charges have been calculated for smaller fragments, like individual amino acids. The so-derived point charges for individual atoms of particular amino acids are then stored in point charge libraries from which they can be retrieved and assigned to each atom in the protein of interest. The often-used AMBER charges, for example, have been determined by scaling point charges to fit the ab initio-derived molecular electrostatic potential [14]. In the case of proteins, the ionization state has also to be taken into consideration. Therefore, formal charges are assigned to those amino acid residues that are expected to exist in charged state under physiological conditions. These charges are placed on one or two of the atoms of a residue. For example, an aspartic acid residue obtains the formal charge −1, which is assumed to be distributed over the two carboxylic oxygen atoms.

In one of the first approaches for a more reliable consideration of electrostatic interaction within proteins, the use of a distance-dependent dielectric constant was introduced. The mathematical equation used for the respective function often has the form $\varepsilon(r) = r$, where r is the distance between the atoms of interest [16]. The distance-dependent dielectric constant is based on plausibility rather than on any experimentally measurable effect. It is assumed that at distances of the order of atomic dimensions the dielectric constant between two charges is that of vacuum conditions, and that at much larger separations the dielectric constant of water $\varepsilon = 80$ holds true. For intermediate distances it is assumed that the dielectric varies with distance in an appropriate way. Distance-dependent dielectric constants can partially mimic the solvent-screening effects on electrostatic energies and are sufficient to stabilize macromolecules in molecular dynamics simulations. However, they cannot correctly describe properties like the electrostatic forces and the electrostatic potential.

A solution of the electrostatic problem may be provided by the use of the Poisson–Boltzmann equation. This equation belongs to the class of differential equations that are typical for the description of boundary phenomena. The Poisson–Boltzmann equation provides a rigorous approach for the calculation of the electrostatic effects of proteins, including the electrostatic potential. Several procedures have been developed which make use of the Poisson–Boltzmann equation. Two commercially available programs are DelPhi [17, 18] and UHBD [10, 19].

In the framework of the Poisson–Boltzman approach the macromolecular system is considered to consist of two separate dielectric regions. The solvent-accessible surface of the protein defines the boundary between these two regions. The interior of this surface is defined as the solute and the exterior is defined as the solvent. Water molecules located in the interior of the protein are usually treated as part of the solute rather than of the solvent. The protein is described in terms of its 3D structure with the location of point charges on the atom centers. A low dielectric constant is used for all points inside the solvent-accessible surface. Common values for this parameter range from 2 to 5. The Poisson–Boltzmann equation is also able to consider the electrostacic effects associated with ions embedded in the solvent. Thus the physiological conditions (0.145 mol l^{-1}) can be incorporated in the calculation.

Use of the Poisson–Boltzmann approach yields the total electrostatic potential of a charged molecule in a solvent according to the following simplified equation:

$$\phi_i^{tot} = \phi_i^{coul} + \phi_i^{self} + \phi_i^{cross} + \phi_i^{own}$$

The solvent molecule responds to the electrostatic field generated by each point charge in the molecule. This response, which consists of two electrostatic effects, the dipolar orientation and the electronic polarization, in turn sets up an electrostatic field at the positions of the original point charges, which is called the *reaction field* [20]. The magnitude of the reaction field is determined by the point charge, its distance from the molecular surface, the shape of the surface and the dielectric constants of molecule interior and solvent. The reaction field exerts a force on all point charges in the system, including the source charge itself. The total electrostatic potential ϕ_i^{tot} is the sum of the interaction of each point charge with its self-reaction field ϕ_i^{self}, the reaction field induced by other point charges ϕ_i^{cross}, the direct coulombic interaction with other point charges ϕ_i^{coul}, and the intrinsic electrostatic potential generated by each point charge ϕ_i^{own} (for a detailed description of this topic, see [5, 8, 21]).

The Poisson–Boltzmann equation is actually a reliable model for the electrostatic interaction in proteins, because it considers the effect of polarization as well as the ionic strength. Unfortunately, it is a very complex differential equation and can be solved analytically only for small regular systems. The alternative to the analytical solution is the use of numerical techniques to find an approximate solution even for large protein systems. For the numerical solution the programs use the so-called *finite difference method* (FDPB). Herein, the protein is mapped onto a 3D cubic grid. The calculated values for the charge density and the electrostatic potential are located on each point of the cubic grid. The numerical solution yields values which are accurate to within 5% in comparison with analytical solutions (which are available for small systems). The most critical regions—and thus the regions of largest errors—are usually those located near charged residues on the protein surface. Several procedures have been recently developed to avoid these errors [18].

The Poisson–Boltzmann method not only offers the possibility to calculate the electrostatic potential of a protein. Additionally, parameters such as the total electrostatic energy of the system, the solvation energy, and the reaction-field energy of proteins can be calculated.

Nevertheless, the most important parameter is the electrostatic potential, which can be displayed in various ways (as described for small molecules in section 2.4.1.2).

Electrostatic potentials have been shown to play an important role in molecular recognition and binding. For example, the electrostatic potential of the superoxide dismutase enzyme has been shown to be responsible for enhanced external diffusion rates of the substrates to the active site [22]. The investigation of the electrostatic potentials of two trypsin enzymes, rat and cow trypsin, has yielded interesting results [23]. These two enzymes, although having the same catalytic mechanism, differ in net charge by 12.5 units. The calculation of the electrostatic potentials, using the Poisson–Boltzmann approach, revealed that both active sites are effectively shielded from the charges located on the surface, resulting in near-identical electrostatic potentials inside the active sites.

As one example for the graphical representation of the electrostatic potential of a protein, gramicidin A, a well-known membrane cation transporting protein, is shown in Fig. 1. Gramicidin A forms a dimer in the membrane. The calculation of the electrostatic potential has been performed for the gramicidin A dimer embedded in a low dielectric membrane layer (which is treated as part of the low dielectric solute system) using the program DelPhi.

4.6.2 Interaction Potentials

Other important features for studying interaction, recognition and binding of possible substrates to a protein are provided by the evaluation of molecular interaction fields. As we have already comprehensively discussed in section 3.2, interaction potentials are useful indicators for the prediction of binding properties of molecules. Programs, like the widely used GRID [24, 25], can be used to map regions within a protein where a water molecule or a substrate is attracted preferentially. The interaction fields, derived with a particular probe, can also be used as a starting point for docking studies of a substrate to its active site. The techniques and procedures applied in this context, are the same as those described in the case of small molecules in section 2.4.2. Various examples are given in literature where these programs have been used successfully to predict binding regions [26–28], to dock molecules into active sites [29–32] and to optimize structures of ligands in order to optimize the binding properties [26, 33, 34].

4.6.3 Hydrophobicity

As we have discussed in section 4.5.2 on the packing quality of proteins, the hydrophobic properties play an important role in the process of protein folding. Also, the protein binding reactivities are often determined by hydrophobic interactions. As was discussed for small molecules (section 2.4.3) several methods are available for the representation of hydrophobic and hydrophilic properties of molecules. The hydrophobicity can be either represented directly on the molecular surface or as a hydrophobic field in the space surrounding the molecule. Useful programs in this respect are, for example, GRID [24], HINT [35] and MOLCAD [36]. A detailed description of the different methods and a comparison of the results derived in studies on proteins is given in the literature [37].

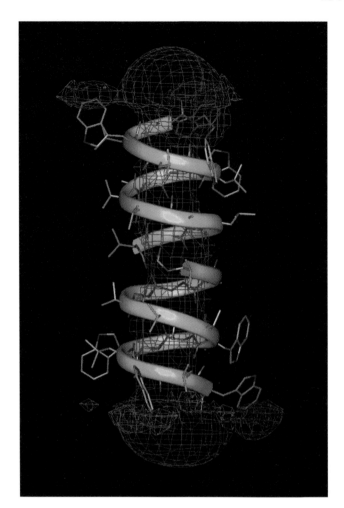

Figure 1. Representation of the electrostatic potential of a gramicidin A dimer embedded in a membrane environment. Calculations were performed using DelPhi. Color code: magenta = negative, green = positive potentials.

References

[1] MacArthur, M. W., Laskowski, R. A., and Thornton, J. M. *Curr. Opin. Struct. Biol.* **4**, 731–737 (1994).

[2] Honig, B, and Hubbel, W. *Annu. Rev. Biophys. Biophys. Chem.* **15**, 163–193 (1986).

[3] Matthew, J. B. *Annu. Rev. Biophys. Biophys. Chem.* **14**, 387–417 (1985).

[4] Warshel, A., and Russel, S. T. *Q. Rev. Biophys.* **17**, 283–422 (1984).

[5] Warshel, A, and Aqvist, J. *Annu. Rev. Biophys. Biophys. Chem.* **20**, 267–298 (1990)

[6] Zauhar, R. J., and Morgan, R. S. *J. Comput. Chem.* **9**, 171–187 (1988).

[7] Gilson, M., Rashin, A., Fine, R., and Honig, B. *J. Mol. Biol.* **183**, 503–516 (1985).

[8] Harvey, S. C. *Proteins Struct. Func Gen.* **5**, 78–92 (1989).

[9] States, D. J., and Karplus, M. *J. Mol. Biol.* **197**, 122–130 (1987).

[10] Davis, M. E., Madura, J. D., Luty, B. A., and McCammon, J. A. *Comput. Phys. Commun.* **62**, 187–197 (1991).

[11] Warwicker, J., and Watson, H. C. *J. Mol. Biol.* **157**, 671–679 (1982).

[12] Warshel, A, and Levitt, M. *J. Mol. Biol.* **103**, 227–249 (1976).

[13] Jorgensen, W. L., and Tirado-Rives, J. *J. Am. Chem. Soc.* **110**, 1657–1666 (1988).

[14] Weiner, P. K., and Kollman, P. A. *J. Comput. Chem.* **2**, 287–299 (1981).

[15] Abraham, R. J., Grant, G. H., Haworth, I. S., and Smith, P. E. *J. Comput.-Aided Mol. Design* **5**, 21–39 (1991).

[16] McCammon, J. A., Wolyness, P. G., and Karplus, M. *Biopolymers* **18**, 927–942 (1979).

[17] DelPhi User Guide, Biosym Technologies, San Diego, California, USA

[18] Gilson, M., Sharp, K., and Honig, B. *J. Comput. Chem.* **9**, 327–335 (1987).

[19] Antosiewicz, J., McCammon, J. A., and Gilson, M. K. *J. Mol. Biol.* **238**, 415–436 (1994).

[20] Bottcher, C. J. F. *Theory of Electric Polarization.* Elsevier Press: Amsterdam 1973.

[21] Gilson, M. K., McCammon, J. A., and Madura, J. D. *J. Comput. Chem.* **9**, 1081–1095 (1995).

[22] Sharp, K., Fine, R., and Honig, B. *Science* **236**, 1460–1463 (1987).

[23] Soman, K. Yang, A., Honig, B., and Fletterick, R. *Biochemistry* **28**, 9918–9926 (1989).

[24] Goodford, P. J. *J. Med. Chem.* **28**, 849–857 (1985).

[25] Wade, R. C., Clark, K. J., and Goodford, P. J. *J. Med. Chem.* **36**, 140–147 (1993).

[26] Reynolds, C. A., Wade, R. C., and Goodford, P. J. *J. Mol. Graphics* **7**, 103–108 (1989).

[27] Wade, R. C., and Goodford, P. J. *Br. J. Pharmacol. Proc. Suppl.* **95**, 588P (1988).

[28] Miranker, A., and Karplus, M. *Proteins Struct. Funct. Genet.* **11**, 29–34 (1991).

[29] Meng, E. C., Shoichet, B. K., and Kuntz, I. D. *J. Comput. Chem.* **13**, 505–524 (1992).

[30] Byberg, J. R., Jorgensen, F. S., Hansen, S., and Hough, E. *Proteins Struct. Funct. Genet.* **12**, 331–338 (1992).

[31] Stoddard, B. L., and Koshland, D. E. *Proc. Natl. Acad. Sci. U.S.A.* **90**, 1146–1153 (1993).

[32] Nero, T. L., Wong, M. G., Oliver, S. W., Iskander, M. N., and Andrews, P. R. *J. Mol. Graphics* **8**, 111–115 (1990).

[33] Varney, M. D., Marzoni, G. P., Palmer, C. L., Deal, J. G., Webber, S., Welsh, K. M., Bacquet, R. J., Bartlet, C. A., Morse, C. A., Booth, C. L., Herrmann, S. M., Howland, E. F., Ward, R. W., and White, J. *J. Med. Chem.* **35**, 663–676 (1992).

[34] Ocain, T. D., Deininger, D. D., Russo, R., Senko, N. A., Katz, A., Kitzen, J. M., Mitchell, R., Oshiro, G., Russo, A., Stupienski, R., and McCaully, R. J. *J. Med. Chem.* **35**, 823–832 (1992).

[35] Kellogg, G. E., Semus, S. F., and Abraham, D. J. *J. Comput.-Aided Mol. Design* **5**, 545–552 (1991).

[36] Heiden, W., Moeckel, G., and Brickmann, J. *J. Comput.-Aided Mol. Design* **7**, 503–514 (1993).

[37] Folkers, G., Merz, A., and Rognan, D. CoMFA as a tool for active site modelling. In: *Trends in QSAR and Molecular Modelling 92.* Wermuth, C. G. (Ed.). ESCOM Science Publishers B. V.: Leiden; 233–244 (1993).

5 Example for the Modeling of Protein–Ligand Complexes: Antigen Presentation by MHC Class I

5.1 Biochemical and Pharmacological Description of the Problem

Cellular immunity is mediated by unique ternary complexes composed of major histocompatibility (MHC)-complex-encoded proteins, antigenic peptides and T lymphocytes. MHC molecules are glycoproteins. Their main function is to bind short antigenic peptides and present them to T lymphocytes at the surface of infected cells (Fig. 1).

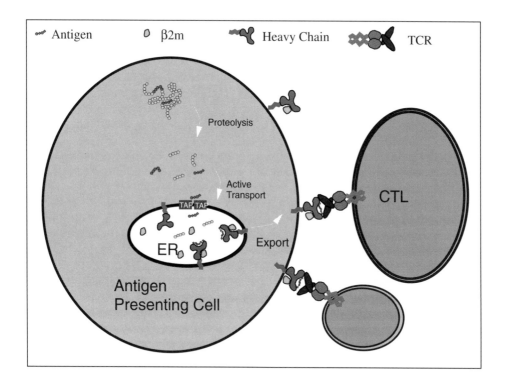

Figure 1. Cellular immune response. CTL, cytolytic T lymphocyte; ER, endoplasmatic reticulum; TCR, T-cell receptor.

5.1.1 Antigenic Proteins are Presented as Nonapeptides

Hence, in contrast to the B lymphocytes, T lymphocytes do not recognize a protein antigen in its native conformation. The protein is normally processed inside the antigen-presenting cell and afterwards brought to the surface and bound to the MHC proteins. The MHC–peptide complex is then recognized by the T-cell receptor on CD8+ T lymphocytes. Most of the antigenic peptides for the class I MHC type are nonamers. This could be shown by elution of the peptides from purified MHC class I molecules. All peptides show conserved residues. Most of them have a conserved amino acid at position 2, which is believed to be the N-terminal anchor residue. Another conserved residue is the C-terminus which is hydrophobic in most cases and sometimes positively charged.

Amino acids at other positions are more variable and make either contact to the T-cell receptor in the ternary complex (T-cell receptor anchor residues) or should not play any decisive role in the formation of the MHC–T-cell interaction complex.

Until now, five MHC class I molecules have been crystallized. They are either bound to a mixture of peptides or to single peptides. Thus, the position of the ligands within the MHC molecules could be unambiguously determined and serves as a basis for the design.

It is a challenge for any design study that the presented antigenic peptides have been shown to be determinative for the whole process of the T-cell response. Length and sequence are the key features for starting the following biological responses:

- assembly and folding of the MHC proteins,
- binding to the MHC molecules,
- transport of this binary MHC–peptide complex to the cell surface,
- recognition of the binary complex by the T-cell receptor.

In terms of a subsequent modeling study, it is important to notice at this point, that:

1. Obviously no empty MHC molecules exist. Therefore, homology modeling of the protein alone does not make sense. This means docking of the ligand and model building of the binding site must take place in an iterative fashion.
2. Binding to the MHC molecule may be achieved by only two residues, namely at positions 2 and 9. This means that criteria have to be found for the discrimination of *good* and *bad* binders, as long as a ternary complex model taking into account the effects of the T-cell receptor, cannot be established yet.

5.1.2 Pharmacological Target: Autoimmune Reactions

Under normal circumstances, the immune system is self-tolerant. However, T-cell receptors which are normally selected to recognize only foreign peptide antigens bound to MHC molecules, may sometimes identify self peptides on MHC class I molecules. Obviously, only the ternary complex and not the MHC complex itself differentiates between self and non-self. T-cell receptors lacking the ability to differentiate between self and non-self may thus break the tolerance of the immune system and cause *autoimmune diseases.*

To date, special forms of arthritis are to our knowledge strongly linked to the expression of certain human leucocyte antigen molecules (MHC molecules). Presentation of bacterial proteins as antigenic peptides which remarkably resemble human self peptides, may be the molecular reason for the autoimmune disease.

In terms of medical treatment of the autoimmune diseases, blocking of the binding site of these special MHC molecules would at first glance be a highly attractive concept for a drug design study.

5.2 Molecular Modeling of the Antigenic Complex Between a Viral Peptide and a Class I MHC Glycoprotein

5.2.1 Modeling of the Ligand

The native ligands of MHC molecules are peptides. At the beginning of a drug design study one starts very often with the description of structural properties of the ligands. This activity is guided by the hope that structure–activity relationships might show up and facilitate the identification of the pharmacophore and/or the docking of the ligand into the binding site.

Peptides however, show considerable flexibility. They have a lot of local energy minima corresponding to a huge variety of different conformations. None of these may be associated with, or relevant for, the bound conformation at the MHC [1]. Furthermore, nothing is known to date about the structural features that determine the antigenic quality of the free peptides. And at last, as revealed by the biochemistry studies mentioned earlier, MHC protein folding seems to be a concerted action process with the binding of the peptidic ligand. The X-ray structures of the MHC complexes showed the bound peptides to have different binding geometries, ranging from an extended state to some sort of coiled geometry.

A set of synthetic peptides derived from the native nonapeptide Tyr-Pro-His-Phe-Met-Pro-Thr-Asn-Leu by subsequent truncation of the N- and C-terminus respectively provided a data basis for a preliminary structure–activity relationship study. A CoMFA study performed with eight peptides, truncated subsequently down from nona- to the pentapeptide, was based on the superimposition of helical geometries of the peptides. The study revealed the importance of the C-terminus to function as an anchor residue [2] (Table 1). The model explains the experimental findings by strong hydrophobic interactions of the C-terminus to a putative hydrophobic binding pocket at the MHC molecule. This information, however, might have been achieved by looking at the isolated C-terminal residues alone. Which relevance had then the helical conformation that had been used for superposition? None !

The helical conformation had been taken because of its local stability. The idea arose from choosing appropriate starting conformations for a dynamic conformational analysis of the peptides. A second clue, that seduced us to accept helical conformations as the most plausible, came from the physical-chemists, who believed, that because of the helix dipole moment this conformation might be the most favored one for establishing protein–ligand interaction.

Both the theory about the importance of helix dipole moments for ligand interaction and the vacuum and solvent dynamics simulation of the isolated peptides, which showed the helical conformation to be the most stable, were found to be wrong in the light of the later

Table 1. Antigenic properties for cytolytic T lymphocytes (clone IE1)

Peptide name	Sequence	Recognition	Peptide concentration
Nona	YPHFMPTNL	+	10^{-9}
Nonar	PHFMPTNLG	+	10^{-3}
Nonal	MYPHFMPTN	–	
Octar	PHFMPTNL	+	10^{-7}
Octal	YPHFMPTN	–	
Heptar	PHFMPTN	+	10^{-7}
Heptal	YPHFMPT	–	
Penta	HFMPT	+	10^{-3}

occurring X-ray structures. However, for reasons of curiosity we had modeled in parallel protein–ligand complexes by energy-minimizing the different peptides bound to the MHC. It evolved that many more than only the helical conformations were preferred in the native environment.

Thus, the important lesson to learn was that peptides as substrates may be handled like other flexible molecules. The binding geometry is strongly case-dependent.

In the present case, some X-ray structures of MHC–ligand complexes, which had been published in the meantime, showed multiple nonapeptides bound to the active site. Their common structural features are two anchor residues. The binding geometry may additionally be markedly influenced by the third binding partner.

Thus, methods like the active analog approach [3], may fail in the case of evaluation of the docking geometries of peptidic ligands, although they have their profound merits in use with synthetic ligands.

This experience led to the decision, to find out as much as possible about the binding site. This knowledge, may it be experimental or theoretical, would help to restrain the degrees of freedom of the peptide's docking geometry.

The aforementioned advent of the first X-ray structures of the MHC class I molecules made it feasible to perform a homology modeling study. Sequences showed more than 70% homology, which should indicate a high degree of structural similarity in that class of proteins.

5.2.2 Homology Modeling of the MHC Protein

Affinity data of the peptides came from the H-2L^d receptor, a MHC-type protein but at present still unknown in structure. A X-ray structure of the human HLA-A2 MHC protein at 2.6 Å resolution was available [4] which shows 70% amino acid homology with the L^d molecule in the a$_1$ and a$_2$ domains (182 residues) of the peptide binding site.

5.2.2.1 Preparation of the Coordinates

In a first step, the crystal coordinates of HLA-A2 were refined in order to remove crystal packing effects. Three different types of calculation were performed with respect to the treatment of electrostatics:

1. A low dielectric model with distance-dependent dielectric functions.
2. A high dielectric model with dielectric constant set to 50 (D50 in Fig. 1).
3. A high dielectric model with explicit water molecules and a dielectric constant set to 1 (DW in Fig. 1).

The "best" structures (based on the deviation from X-ray backbone structure) were obtained by use of the high dielectric models. The models with the distance-dependent dielectric function very often overestimated salt bridges for instance between lysines and acidic amino acids, thereby creating non-regular structures. Thus, for a subsequent molecular dynamics simulation the distance dielectric model was dropped and only the "good" high dielectric models were used.

The molecular dynamics simulation procedures produced major discrepancies between the two starting structures of the model. The model with the dielectric constant set to 50 produced an unacceptably large deviation of almost 4 Å compared with 2 Å deviation obtained in the model with the explicit water molecules. Therefore, only the latter was found to provide a realistic picture at least near the solid-state geometry in the crystal and with minimized internal energy.

When inspecting the details of structure deformations in the vacuum dynamics simulation (model with dielectric constant set to 50) two prominent features could be seen to be responsible for the large rms deviation and typically occurring in vacuum simulation. Firstly, the active site, aligned by the two large helices shrank considerably by more than 50% (Fig. 1). Secondly, the helices themselves shrank by up to 6 Å. From this, a binding site resulted that would never be able to accomodate any ligand and thus was worthless for any further drug design procedure.

Part of the phenomenon can be explained by artificial hydrophobic collapses occurring with in vacuo simulations, because the hydrophobic surface attemps to become minimal. In contrast, by using of the explicit water model, the structure was seen to fluctuate around the X-ray-defined structure, giving the side chains the possibility of finding the optimal energy level with respect to a solvent environment. This led to a structural model averaged from 150 single structures during molecular dynamics simulation, that showed a backbone topology very near that of the X-ray backbone. Both the X-ray structure and its refined model by explicit treatment in a solvent are able to accomodate a nonapeptide as ligand in their binding site.

5.2.2.2 Building the H-2L^d Molecule

The coordinates minimized in the explicit solvent environment were taken as a basis for the construction of the homology model of H-2L^d from HLA-A2. During the procedure, only the side chains were modified; the backbone was kept untouched. As has been described earlier, side chains were exchanged in a first step without taking care of interactions or optimal electrostatics. Because of differences in the sequence, a deletion occurred near the N-terminus.

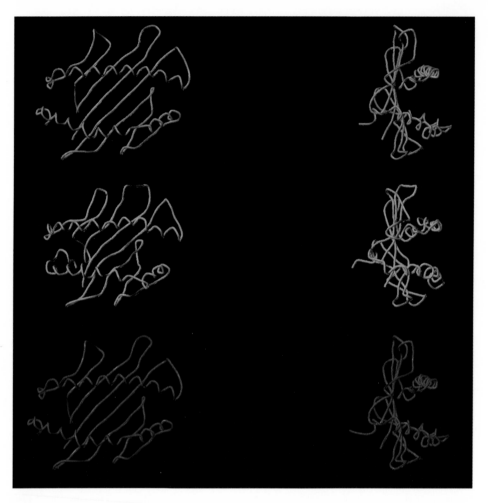

Figure 1. Orthogonal views of three HLA backbone conformations: X-ray (top), D50 mean conformation (middle) and DW mean conformation (bottom).

This deletion was located in a loop structure. The latter is to be expected because at such a level of homology of both of the sequences, helices and sheets are always conserved. However, connecting loops are the positions, where individual substitutions occur in order to accomodate evolutionary fitting processes between different tissues or different species.

The loop identified between residues 12 and 18 had to be reconstructed from scratch. Because loops often have no ordered structure—or assume ordered structures only in the presence of a binding partner—we decided to perform a "loop search" in the Brookhaven crystallographic database in order to obtain an at least acceptable structure of the newly built loop. "Loop searches" perform a sequence alignment of the sequence to be built with the sequences already present in the protein database (Fig. 2).

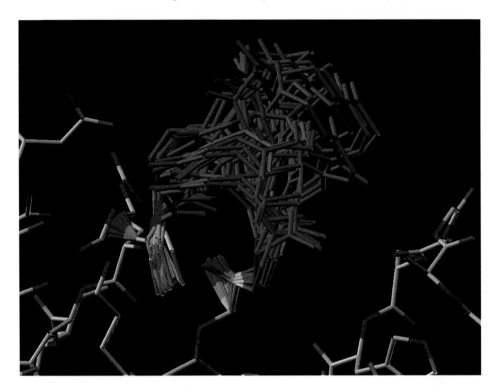

Figure 2. Building a loop using the SYBYL_ loop search algorithm. Proposed backbone conformations are shown in red.

The algorithm is contained in most of the leading modeling packages. It presents the ten best "matches" from sequence comparison and subsequently cuts out the respective loop structures from the protein X-ray structures. The best fitting loop can be choosen to be built into the homology model, with respect to the distance of the N-terminal and C-terminal (see also section 4.3.3).

In the present case we found a loop with moderate homology but having a backbone geometry with only a 0.38 Å rms deviation from the N–C terminal distance defined by the template structure. At this stage the homology model represents a rough assembly of side chain orientations that must be refined in the subsequent steps.

The question is, whether this must be done in the presence of a docked ligand or with an "empty" binding site. According to the literature, as well as to our own experience, refinement of the complex should be performed preferentially with the ligand bound to the protein. However, the early steps might be done without any ligand, because the disorder of side chains may be too large to dock a ligand straight away in the binding site. The situation is worse in proteins that show an induced fit; in these cases multiple homology modeling steps are needed.

In the present case, we were sure that the MHC molecule was only folded correctly in the presence of the ligand. Therefore, we kept the backbone constant and removed steric interaction by energy-minimization. Subsequently the homology model was subjected to molecular dynamics simulation because we were curious to see whether it behaved like the X-ray structure; Indeed it did. Again, the model with the explicit water treatment showed a result much closer to the X-ray backbone, than to the model with dielectric constant set to 50. This analogy to the behavior of the X-ray structure gave us some confidence that the homology model obviously possesses at least some protein properties

As mentioned earlier, MHC molecules usually fold only in the presence of ligands; this led us to attempt a peptide docking in order to achieve the whole binary interaction complex and to proceed with the structure refinement of the complex. From previous QSAR studies it was suggested that the C-terminus of the ligand should bind to a hydrophobic environment or pocket. The peptides showed that at least positions 1 and 2 should additionally contribute to binding to the MHC molecule. We began to look for a binding pocket with hydrophobic properties that could accomodate the C-terminal amino acid of the peptide ligand and was limited in size, so as to exclude the amino acid tryptophan, which caused inactivity in the biological tests.

As a graphical aid, we used a hydrophobicity coloring scheme for the surface of our homology model of the H-2L^d molecule. The Fauchère–Pliška scale was applied to characterize hydrophobicity [5] (see Fig. 3). This particular scale was chosen because it has

Figure 3. Space-filling representation of the H-2L^d molecule with the Fauchère–Pliška hydrophobic scores. Color scheme: hydrophobic, magenta and cyan; hydrophilic, yellow and red.

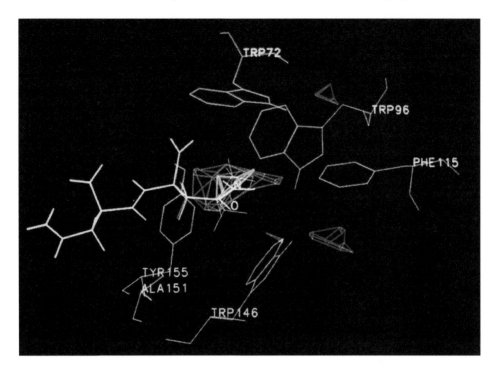

Figure 4. Part of the binding pocket of the HLA C-terminus.

been determined experimentally and has already been successfully applied in studying antigenic sites in proteins. As expected, hydrophilic residues are located mainly at the surface of the protein, whereas hydrophobic areas are buried. One hydrophobic pocket, however, seemed quite large and suitably sized for the docking of the C-terminus. It consisted of three tryptophan, two phenylalanine and two tyrosine residues (Fig. 4).

Further indications came from experiments; two X-ray studies showed extra electron densities at the position of the hydrophobic pocket, supposedly resulting from co-crystallized peptides. The resolution, however, was not good enough to detail the interactions. This prompted us to choose the described pocket at the site for the C-terminal amino acid to serve as an anchor site for the peptide ligands. What of the remainder of the ligand's geometry? Residues at positions 1 and 2 had been predicted by QSAR and biochemically to be important also for the ligand's interaction with the MHC molecule, but nothing was known of the rest.

The helical interactions of the peptide ligand

Unfortunately we were seduced by the results of our conformational studies on the ligand. As mentioned earlier an α-helical structure was given to the ligands based on 3D–QSAR studies and for two additional reasons.

1. The α-helix turned out to be the most stable structure in solvent as predicted by molecular dynamics. However, this is the wrong line of enquiry. Minimization inside the binding site only provides useful information about the ligand, not energy minimization in vacuo or even in solvent. Nevertheless, we docked the ligand with helical geometry into the H-2L^d binding site. None of the currently known X-ray structures of MHC–ligand complexes show a regular helical conformation of the bound peptide. Therefore it is not likely that this was the only case of ligands being bound in such geometry. Strangely, manual docking provided an excellent interaction geometry for the nonapeptide Tyr-Pro-His-Phe-Met-Pro-Thr-Asn-Leu. The Leu9 fitted nicely into the hydrophobic pocket (Fig. 5(c)). The N-terminal Tyr1 interacted by aromatic interaction with Trp166 and by electrostatic interaction with Tyr56 of the MHC molecule (Fig. 5(a)). The second residue Pro2 could also be located nicely adjacent to Leu62 and Leu65 of the MHC, showing perfect match of the solvent accessible area and hydrophobic interactions (Fig. 5(b)). Furthermore, the α-helix ideally spans the space between C- and N-terminal docking position.

2. Further support for this docking alternative was given by the fact, that in the present model, the nonapeptide interacted with highly polymorphic positions of MHC proteins, which could be interpreted as being a specific interaction. Of special note was the Ile62 and Ile65 contacting the proline at position 2, which is unique in these molecules.

This was a very optimistic view of the docking and ligand interactions. The opposite interpretation would be also possible, and is probably the more realistic. Although specificity is needed for ligand–MHC interaction, the main contacts are made to conserved residues in the class I MHC molecules. Therefore, the helical conformation that causes interactions of the ligand to non-conserved residues may be incorrect.

One may criticize that *any* model of antigen recognition must necessarily be incomplete as long as the contribution of the T-cell receptor (TCR) in the ternary MHC–peptide–TCR complex cannot be included in the simulation. Yet, to this argument, the fact must be recalled that the formation of the ternary complex is a stepwise process. First, the peptide must bind before the now-formed binary complex is subsequently recognized by the T-cell receptor. This again demonstrates the importance of integration of biochemical knowledge in the modeling process. Nevertheless we do not know, until the X-ray structure or model of the ternary complex with the T-cell receptor is available, which geometry is the real one. In conclusion, the resolution of the model was inadequate to provide a unique geometry for the protein–ligand and interaction complex.

The extended non-regular interaction of the peptide ligand

In a parallel study we examined the interaction of a peptide derived from influenza virus protein with the HLA-A2 MHC molecule that has been used as a template for the homology model of H-2L^d in the previous section [6].

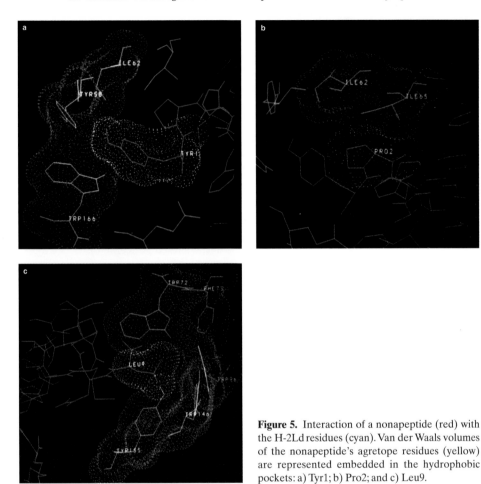

Figure 5. Interaction of a nonapeptide (red) with the H-2Ld residues (cyan). Van der Waals volumes of the nonapeptide's agretope residues (yellow) are represented embedded in the hydrophobic pockets: a) Tyr1; b) Pro2; and c) Leu9.

Before docking the nonapeptide Gly-Ile-Leu-Gly-Phe-Val-Thr-Leu into the binding cleft of the HLA-A2 X-ray structure, the structure was truncated to the a_1 and a_2 domains. This approximation has been shown before not to alter significantly the 3D structure of a_1/a_2 because only limited interactions exists between a_1–a_2 domains and the a_3 and β-microglobulin domains. The latter are not suspected to contact antigenic peptides. The C-terminus was protected by a N-methyl group in order to avoid unrealistic electrostatic interactions. Furthermore, three crystalline water molecules have been placed in the binding site because they are visible in the X-ray structure and may be of importance for support of the peptide binding.

Subsequently, the nonapeptide Gly-Ile-Leu-Gly-Phe-Val-Phe-Thr-Leu was docked manually in the peptide binding groove. This time we used as a "template" an extra electron density map located in the binding groove of the MHC molecule that was observed in the X-ray structure, but could not be resolved to a unique ligand.

The crystallization of a mixture of peptides with the HLA-A2 molecule might have caused this extra electron density. Nevertheless, it provided a volume restraint into which the ligand should fit (Fig. 6). The N-terminal glycine was fixed at a hydrogen bonding distance from the conserved residues Tyr7, Tyr59 and Tyr171 (Fig. 7). The second residue, isoleucine, is one of the formerly detected anchor residues, conserved among the peptidic ligands of MHC molecules. It has been placed such that its hydrophobic side chain contacted a set of three valines that form part of a hydrophobic pocket—which is also conserved among the MHC molecules. The peptide was further extended to the third and fourth residue by fitting it to the electron density map. Ile3 pointed in direction of a pocket named D, which was also hydrophobic in character aligned by two tyrosines and two leucines. Ile3 did not fill this pocket completely.

Again, this is a branching point in a modeling study that gives rise to two interpretations. Firstly, if there is a pocket it should be filled completely by the ligand's side chain in order to avoid "empty space" or large entropic contribution. It is unlikely that in the hydrophobic pocket water molecules fill the empty space. Secondly, the contrary argument is that evolution favors an optimal solution, not a maximal one. The ligand should be able to dissociate again. Filling every binding pocket to a maximum would considerably increase the energy necessary for dissociation and would decrease the variety in recognition of other peptides without losing specificity. Moreover, by an ultra high-affinity, a specificity would be gained that is probably not needed or that would even prevent the immune response.

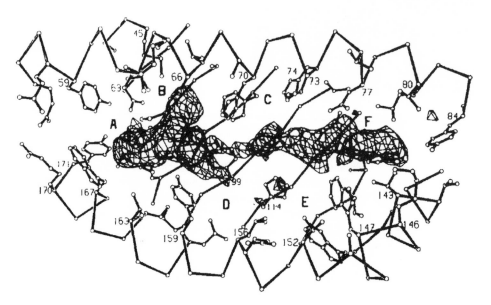

Figure 6. Extra electron density in the X-ray structure of the HLA-A2 co-crystallized with a mixture of nonapeptides.

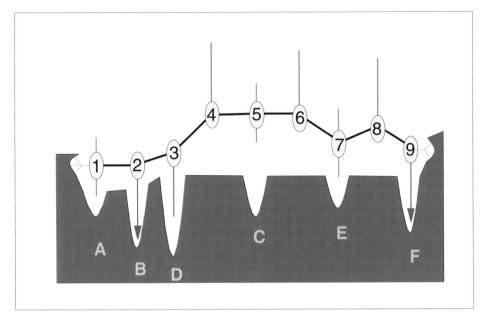

Figure 7. Simplified model of nonapeptide binding to class I MHC proteins exemplified by HLA-B27. The six specificity pockets are labelled from A to F. N- and C-termini are colored in magenta. Main anchor side chains at P2 and P9 are displayed as arrows. MHC-binding side chains are colored in green, potential T-cell receptor (TCR)-binding side chains in red. Some amino acids (side chains in cyan) may bind both MHC and TCR molecules.

Experimental constraints at branching points

At branching points the modeler needs help in deciding which branch to follow. Very often this information can be taken from previously known biochemical and pharmacological data. Therefore it is very important, as discussed earlier, to store and use all experimental information that can be accessed concerning the target protein and the ligands. In the present case the branching point provides an excellent opportunity to design a biochemical experiment that will prove the modeling process and the prediction, respectively.

Thus, we decided to design a non-natural peptide with maximum interaction at pocket D. From the synthetic ligand, we expected a much higher affinity to the MHC. (The design step will be described in detail in the next section.) In consequence we proceeded to model the natural ligands along the extra electron density template and let the pocket be only partly filled by Ile3. From there, the backbone turns upwards, directed to the opening of the binding cleft pointing to the solvent. For Gly at position 4 and Phe at position 5 a solvent or T-cell receptor interaction is hereby assumed. Val6, in contrast, is again contacting the MHC at a binding pocket, that is surprisingly polar, formed by two histidines, a threonine, and an arginine.

Positions 7 and 8 of the peptide ligand, following the electron density template, again point in the direction of the solvent. Interestingly, this is supported by a high variability at this position for the peptides eluted from the MHC complex in biochemical experiments.

Finally, position 9, bearing a leucine, was docked into the well-known hydrophobic C-terminal pocket, in line with details in the previous section regarding the H-2L^d molecule. This docking alternative created no steric conflicts and seemed rather reasonable from a chemical viewpoint. Although having a quite irregular, extended geometry, no violations of angles, torsion, etc. were identified on visual inspection of the interaction complex.

5.3 Molecular Dynamics Studies of MHC–Peptide Complexes

Modeling and docking of ligands has been described for H-2L^d and HLA-A2. In the next section the molecular dynamics simulations of two complexes will be discussed, namely HLA-A2 and—far more interesting from a viewpoint of unexpected results—HLA-B*2705 [6].

5.3.1 HLA-A2—The Fate of the Complex during Molecular Dynamics Simulations

For the HLA-A2 system, the homology modeling described in the previous section, the bimolecular complex, and the three crystalline water molecules were placed in a shell of water molecules. No periodic boundaries were applied, nor were positional constraints placed on the solvent atoms.

As usual, the solvated complex was minimized first and subsequently subjected to a 100 ps molecular dynamics simulation at constant temperature. The system was coupled to a heating bath (see earlier). Analyses were taken from the last 60 ps. From the period of analysis (40–100 ps) a mean structure of the molecular dynamics simulation was obtained by averaging the atomic coordinates for 600 conformations. As shown in Fig. 1 the overall geometry is unchanged compared with the X-ray structure during molecular dynamics situation. The docking of the peptide did not significantly change the HLA-A2 structure. Most of the structural deviation—or even artefacts—were observed in loops connecting the α and β structures and the β-sheet, respectively. The more severe artefact was an unexpected flexibility of the a$_2$ domain. This could be easily explained due to the lack of following a$_3$ and b$_2$ domains which are the native constituents of the complete MHC molecule, and are not present in the model. The general folding, however, was not disturbed because this movement was found oscillating around the X-ray structure.

A more detailed inspection of potential structure distortion was performed by using selected fits of secondary structures. This allows for distinguishing rigid body motions from distortion. In the first case, a structural element—for instance a helix—is translated or rotated as a whole. In a global fit, this would cause a bad rms value because of the average over the whole structure. If, however, only the structural element is fitted, it can be seen immediately that if the rms drops drastically, than the geometry is maintained and only the structure element has been translated or rotated. If, however, the rms value increases, a distortion of geometry is indicated.

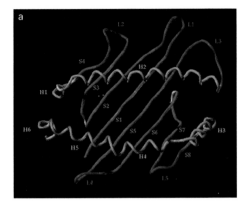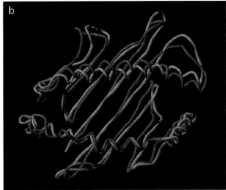

Figure 1. Experimental and simulated 3D structures of HLA-A2.1. a) Crystal structure of the MHC protein. Only a1 and a2 domains are shown (α-helices H1–H6: yellow; β-strands S1–S8: green; loops L1–L5: red; water: cyan balls). b) Superimposition of the crystal structure (cyan) and the time-averaged simulated conformation (red).

Looking at the HLA-A2 step-wise by comparing secondary structure elements (Table 1) all but three secondary structure elements showed a considerably reduced rms, indicating that the overall geometry was maintained. The three elements that were more affected and showed an increased rms value are located near the C-terminus of the model. As explained earlier, the model takes into account only the binding site, consisting of a_1 and b_2 domains. The lacking a_3 and b_2 domains definitely do not contact any antigen, but might stabilize the whole MHC molecule. Therefore, if these domains are lacking just adjacent to a_2, some higher flexibility would be expected, probably resulting in distortion—exactly the situation that we found.

In an even more detailed step we traced the backbone angle variation. Around 80% of the ϕ and ψ angles did not alter for more than 20°. Only in the loop region were larger deviations found and, indeed, would have been expected. Interestingly, larger ϕ and ψ deviation which occurred at the C-terminal elements were always compensated by the next ϕ and ψ angles along the sequence, thus maintaining perfectly the overall secondary structure and the interaction geometry.

To analyze observations and statistics of atomic fluctuations is again one level deeper in a detailed study of the system. All fluctuations correlated with the motions of larger parts and substructures described earlier.

Atomic fluctuation has been analyzed in the present case, especially for water molecules, the whole system being surrounded by some 1300 water molecules with TIP3P potentials.

Consideration of atomic fluctuation of water oxygen atoms is one method of analyzing the quality of the molecular dynamics simulation. We could clearly detect four different types of water: first, the explicit water molecules inside the binding cleft showing participation in the hydrogen network of the ligand; second, water molecules bound to the surface of the interaction complex, exhibiting fluctuations only; third, the bulk water with moderate variation; and fourth, water molecules at the water–vacuum edge being the most flexible and showing rms values over 1.0

Table 1. Root mean square (rms) deviation from the crystal structure. Coordinates of the HLA-A2 a1–a2 domains were time-averaged over 40–100 ps and compared with the crystal structure. Deviations in nm have been calculated for backbone atoms after fitting the whole structure (rms 1) or selected sequences (rms 2)

Structure	Position	rms1	rms2
a1–a2 Domains	1–182	0.182	0.182
a1 Domain	1–90	0.167	0.162
a2 Domain	91–182	0.195	0.191
secondary elements		0.166	0.078
α-Helices		0.173	0.091
H1	50–53	0.206	0.128
H2	57–84	0.149	0.099
H3	138–150	0.138	0.033
H4	152–161	0.091	0.043
H5	163–174	0.211	0.137
H6	176–179	0.371	0.042
β-Strands		0.120	0.066
S1	3–12	0.074	0.074
S2	21–28	0.073	0.043
S3	31–37	0.083	0.043
S4	46-47	0.122	0.024
S5	94–103	0.076	0.056
S6	109–118	0.076	0.122
S7	121–126	0.229	0.074
S8	133–135	0.185	0.044
Loops		0.250	0.106
L1	13–20	0.226	0.137
L2	38–45	0.264	0.128
L3	85–93	0.175	0.131
L4	104–108	0.348	0.024
L5	127–132	0.201	0.057
Crystal water	193–195	0.249	0.132

5.3.2 HLA–B*2705

For the HLA-B*2705 MHC molecule a X-ray crystal structure was available [7] from which the coordinates were taken. The peptides bound to the MHC molecules were derived from the nonamer which had been modeled into the binding cleft of the X-ray structure; this had the sequence Ala-Arg-Ala-Ala-Ala-Ala-Ala-Ala-Ala. The other peptides have been created simply by replacing the alanines subsequently by the corresponding residue of the desired derivative. Its side chains were centered in the binding-pockets according to the electron density map of the peptide. Those residues responsible for the interaction with the T-cell receptor pointed towards the solvent as no receptor interaction could be taken into account. The backbone geometry of the crystal structure was taken as a template for all nonapeptides bound to the MHC [8].

So far, the entire situation is quite similar to the complex described earlier, so, what is interesting about B*2705? The B*2705 binding motif has been characterized by analysis of a variety of the bound peptides. These were eluted from the native complex by HPLC and sequenced. From the set of peptides available, position 2 was identified as a main anchor residue, always being an arginine. The other anchor positions are 1, 3, and 9, preferring hydrophobic and positively charged residues, and 2 and 9 being the most important. These experimental data do not, however, entirely account for the HLA-B*2705 binding properties of several bacterial peptides.

Peptides from *Chlamydia trachomatis* could be shown to bind to HLA-B*2705. They stem from the 57 kDa heat shock protein of *C. trachomatis* and are probably responsible for an anti-immune reaction causing diseases related to rheumatoid arthritis. The bacterial peptide Leu-Arg-Asp-Ala-Tyr-Thr-Asp-Met-Leu, for example, fits nicely the consensus sequence. Arginine in position 2 and a hydrophobic or positively charged amino acid in positions 1, 3, and 9 represent a binding pattern as defined by the anchor residues, except for position 3. Thus, the peptide is expected to show affinity to HLA-B*2705, but does not because it is not recognized. The opposite is true for the peptide Arg-Arg-Lys-Ala-Met-Phe-Glu-Asp, i.e. an octapeptide rather than a nonapeptide, but with the only similarity being the arginine in position 2. If this position is correctly docked into the second pocket, the peptide would be too short for any interaction with the hydrophobic pocket at position 9, this being very important for stabilization of the nonapeptides. Surprisingly, the octapeptide is recognized by the MHC molecule, although its binding motif does not match the experimental pattern very well.

These were the reasons why we were interested in the B*2705 complex and tried to rationalize the structure–activity relationships by performing molecular dynamics simulations. Six different peptides were chosen for this study, the rationale for choice being the following (see Table 2):

1. The nonapeptide Arg-Arg-Ile-Lys-Ala-Ile-Thre-Leu-Lys has been described as part of the crystal structure. Therefore this peptide served as a basis for the setup of appropriate parameters for the molecular dynamics simulations. If the X-ray structure could be reproduced by a certain set of molecular dynamics parameters, we would be willing to accept these conditions for the whole series of peptide–HLA complexes to be simulated. We are fully aware, that this assumption is in fact an extrapolation. It is, however, the most cautious one that can be done in this case.

2. The second (Glu-Arg-Leu-Lys-Glu-Ala-Ala-Glu-Lys) and third (Arg-Arg-Lys-Ala-Met-Phe-Glu-Asp-Ile) peptides were taken as positive controls, because they showed high-affinity binding. The sequence Glu-Arg-Leu-Ala-Lys-Leu-Ser-Gly-Gly has been taken as a negative control, as it does not bind to the B*2705.

Both peptides mentioned previously (Arg-Arg-Lys-Ala-Met-Phe-Glu-Asp and Leu-Arg-Asp-Ala-Tyr-Thr-Asp-Met-Leu) were used as test cases, where we hoped to be able to explain the unexpected binding properties.

For the docking of the octapeptide we had to accept some compromises. The negatively charged Asp at the C-terminus was certainly not expected to interact with the pocket for the normal C-terminal residues of the nonapeptides because the pocket itself is aligned by two

Table 2. Binding of bacterial peptides to HLA-B*2705

Peptide no.	Sequence	Origin	Binding
1	Arg-Arg-Ile-Lys-Ala-Ile-Thr-Leu-Lys	Model	not determined
2	Gln-Arg-Leu-Lys-Glu-Ala-Ala-Glu-Lys	Hsp 75[a]	good
3	Arg-Arg-Lys-Ala-Met-Phe-Glu-Asp-Ile	Hsp 57[b]	best
4	Glu-Arg-Leu-Ala-Lys-Leu-Ser-Gly-Gly	Hsp 57	non
5	Leu-Arg-Asp-Ala-Tyr-Thr-Asp-Met-Leu	Hsp 57	non
6	Arg-Arg-Lys-Ala-Met-Phe-Glu-Asp	Hsp 57	good

[a] From *Escherichia coli*.
[b] From *Chlamydia trachomatis*.

Asp residues. Thus, the docking was based on the hypothesis that the Asp might be able to simulate the normal C-terminus of the nonapeptide. Therefore, the octapeptide was docked without having a side chain interaction of pocket F, which is normally responsible for binding the C-terminus of the nonapeptide. The more extended conformation of the octapeptide could be accomplished by moderating the bulge, normally occurring between position 4 and 7 and in reality supposedly binding to the T-cell receptor (Fig. 2).

5.3.2.1 The Fate of the Complex during Molecular Dynamics Simulations

Here we describe only the main concepts used to distinguish between "good" and "bad" binders. The detailed analysis with listings of every hydrogen bond interaction may be duplicated in the original papers [8, 9].

In fact, the molecular dynamics simulation proved able to account for anomalous binding of the bacterial peptides. Again, as shown previously, the most important criteria for the judgment of the models were hydrogen bonding, solvent-accessible areas, and atomic fluctuations. To begin with the latter, we were mainly interested in the behavior of the binding pockets related to anchor residues 2 and 9. When bound to the inactive peptides Glu-Arg-Leu-Ala-Lys-Leu-Ser-Gly-Gly and Leu-Arg-Asp-Ala-Tyr-Thr-Asp-Met-Leu, respectively, atomic fluctuations were increased dramatically compared with the native peptides. As expected, the atomic motions of the pockets correlate clearly with the nature of side chain of the peptides. Good binders have perfectly complementary side chains properties. Thus, inactive peptides lack side chain interactions or show only weak interaction with pockets 2 and 9. This results in an increased atomic mobility. Logically, a similar pattern emerged for the analysis of hydrogen bonding in the peptide–MHC interaction. Again, our main interest focused on the binding pockets for residues 2 and 9.

▶

Figure 2. Time-averaged conformation of HLA-B27 in complex with six peptides (A–F). The backbone atoms of the two α-helices delimiting the peptide-binding groove are displayed here with the side chains of peptide-binding residues. The C^2 positions of bound peptides (in bold) are labeled from P1 (N-terminus) to P9 (C-terminus). Only the peptide anchor side chains are shown. MHC–peptide hydrogen bonds are represented by broken lines and water molecules by balls.

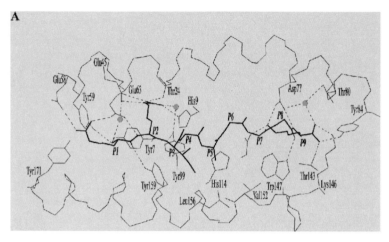

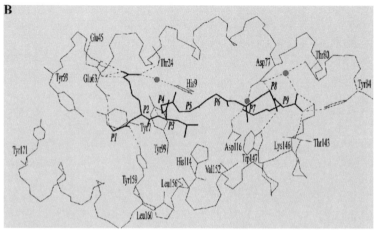

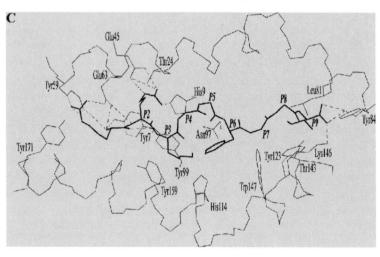

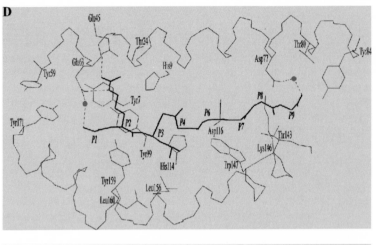

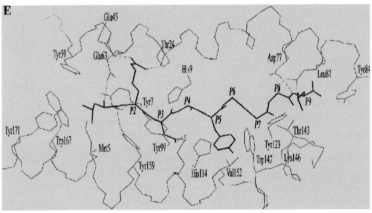

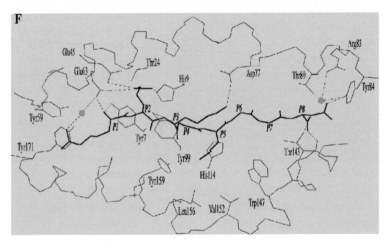

For the native complex (X-ray structure) from 15 hydrogen bonds interactions at positions 2 and 9, all but four could be reproduced by the molecular dynamics simulation. This is quite a lot if one considers the highly reductionistic model. Interestingly, these four missing H bond interactions compared with crystal structure could be shown to be engaged in the water intercalation effect at the N-terminus. Water molecules slowly "walk in", starting at the N-terminus, slightly loosening the side chains from their binding pockets. Although this might be an artefact of the simulation, the principle reflects the differences between solution and crystal state.

A dramatic drop of H bond interactions is seen for the inactive peptides; this was also expected from atomic fluctuation analysis with only two of seven H-bonds remaining. For both peptides, the C-termini have lost their original H-bonds, while at the N-terminus the peptides are not hydrogen bonded at all.

The most interesting situation is that for the octapeptide. Thirteen H bonds emerge after and during 150 ps molecular dynamics simulation. The main anchor residue arginine resides in pocket B, its native location. The middle part of the peptide (residues 4–7) does not interact at all with the MHC molecule. The C-terminus, however, is in fact replacing the normal carboxyl end of the nonapeptide (Fig. 2).

5.4 Analysis of Models that Emerged from Molecular Dynamics Simulations

Four criteria have been used to analyze the binding situation of ligand–protein complexes and to correlate them at least non-quantitatively to the experimental observations. The criteria were: hydrogen bonding networks, interaction energies, solvent-accessible surface, and atomic fluctuations.

Attempts to quantify the results of the molecular dynamics simulation were very difficult. Therefore, the use of calculated interaction energies may be the weakest part of the four criteria mentioned. Simplification in quantification of electrostatic interaction and hydrophobic binding, respectively, will provide only rule-of-thumb values for estimation of ligand–protein interactions. Only in rare cases does the reductionistic nature of the models allow for a *quantitive* structure–activity relationship based on interaction energies. Thermodynamic analyses of ligand–protein interaction are currently under study and may be used in future to calibrate calculated interaction energies. Furthermore, refined approaches to calculate electrostatics—as designed by the use of the Boltzmann–Poisson equation [10]— may be helpful in the detailed quantitative analysis of interaction energies.

5.4.1 Hydrogen Bonding Network

In the studies presented, hydrogen bond properties have been described in terms of a donor (D)–acceptor (A) distance lower than 0.35 nm and a D–H–A bond angle value of 120–180°. Time-averaged conformations of up to 200 ps simulation time were analyzed in most cases.

In general, the total number of MHC–peptide hydrogen bonds was strictly correlated to the binding properties of the corresponding peptide. In all simulations the significant pattern of hydrogen bonding networks could be reproduced for crystal structures. For the non-binders—peptides exhibiting low experimental affinities—a dramatic loss of hydrogen bonds could generally be observed. This was especially true for the N- and C-termini. The effect was less dramatic for the main anchor position 2.

Differences in the hydrogen bond pattern, when compared with the crystal structures, may also occur with high-affinity ligands, for example, water molecules moving in at C- and N-terminal binding pockets, though whether this is an artefact or simply shows an early step of dissociation remains unclear.

In our opinion molecular dynamics simulations represent much of a solvated state. Furthermore, the molecules are provided with kinetic energy which enables them to find new positions, not necessarily tracing down to the global minimum. Thus, differences to the crystal state might be expected which—if they occur coincidentally at the termini of the bound peptides—may cause the molecular dynamics simulations to reflect something of the reality of dissociation behavior of the ligand–protein complexes. In the present case of ligand–MHC interaction, careful analysis of the hydrogen bonding pattern was the basis for predicting correctly these parts of the ligands that could be replaced by non-interacting spacer residues (see the next section).

Table 1 shows how such a H-bond pattern emerging from the molecular dynamics simulation can be represented. Low-affinity binders (peptides 4 and 5) can be detected directly by loss of H-bond interactions compared with the crystal structure of a native ligand bound to the MHC (first column).

5.4.2 Atomic Fluctuations

The atomic fluctuations were computed and compared with the fluctuations from the crystallographically determined temperature factors. This allows for an illustration of the relative gain or loss in flexibility compared with the native X-ray structures.

Atomic fluctuations were found to be an excellent tool which provides direct insight in the activity-correlated properties of the ligands, as they depend directly on strong or weak electrostatic and/or hydrophobic interactions. Graphical representation facilitates the direct comparison of several ligands with respect to their binding properties within the same scale. This is illustrated by the following example (Fig. 1).

The upper graph in Fig. 1 represents the atomic fluctuations of the binding pockets. Fluctuations are calculated from time-averaged conformations of the last 500 conformations of the molecular dynamics simulation. The binding pockets themselves are formed by up to six side chains. If one looks at the active peptides, which means high-affinity binders, pockets B for residue 2, pocket D for residue 3, and pocket F for residue 9, show the lowest fluctuations. The second amino acid of every peptide is bound to pocket B with such a high affinity, that the movement of the side chains of the pockets is dramatically restrained. This is represented by the solid lines clustering around 0.65 Å rms fluctuations.

Switching to the low-affinity peptides (the non-binders), the situation changes completely. The residues forming pocket B show rms fluctuations between 0.7 and 0.8 Å. This indicates a larger flexibility of the binding pocket and, vice versa, a less tight binding or only few interactions

Table 1. MHC-peptide hydrogen bonds. Peptide positions (Pn) are labelled from 1 (N-terminus) to 9 (C-terminus). Closed and open boxes indicate the presence or absence of a peptide–MHC hydrogen bond, respectively (time-averaged distance between donor D and acceptor A less than 3.2 Å, D–H⋯A angle between 140° and 180°). Crosses indicate the absence of specific side chains for some peptides. Hydrogen bonds have been analyzed for the crystal structure (X-ray) and during the last 50 ps of the simulation over 500 HLA-peptide conformations for each complex with peptides 1 to 6 (MD1 to MD6)

Peptide	HLA*B2705	X-ray	MD1	MD2	MD3	MD4	MD5	MD6
P1(N)	Tyr7(OH)							
	Tyr59(OH)							
	Glu63(OE1)							
	Glu63(OE2)							
	Tyr171(OH)							
P1(NE)	Glu163(OE2)			X		X	X	
P1(NH1)	Glu63(OE2)			X		X	X	
P1(NH2)	Glu58(OE2)			X		X	X	
	Glu166(OE2)			X		X	X	
P1(O)	Tyr99(OH)							
	Tyr159(OH)							
P2(N)	Glu63(OE1)							
P2(NE)	Glu45(OE1)							
	Glu45(OE2)							
	Glu63(OE2)							
P2(NH1)	His9(NE2)							
	Thr24(OG1)							
	Glu45(OE1)							
P2(NH2)	His9(NE2)							
	Glu45(OE1)							
	Glu45(OE2)							
P2(O)	Arg62(NH1)							
P3(N)	Tyr99(OH)							
P3(OD2)	Gln155(NE2)	X	X	X	X	X		X
P3(NZ)	Asp77(OD1)	X	X	X	X	X	X	
P8(O)	Lys146(NZ)							
	Trp147(NE1)							
P8(OXT))	Thr143(OG1)	X	X	X	X	X	X	
P8(OD2)	Tyr84(OH)	X	X	X	X	X	X	
P9(N)	Asp77(OD1)							X
P9(NZ)	Asp77(OD2)				X	X	X	X
	Asp116(OD2)				X	X	X	X
P9(O)	Tyr84(OH)							X
	Thr143(OG1)							X
	Lys146(NZ)							X
P9(OXT)	Tyr84(OH)							X
	Thr143(OG1)							X
	Lys146(NZ)							X
Total		15	18	16	13	2	7	13
Backbone		10	9	9	8	1	4	6
Side chains		5	9	7	5	1	3	7

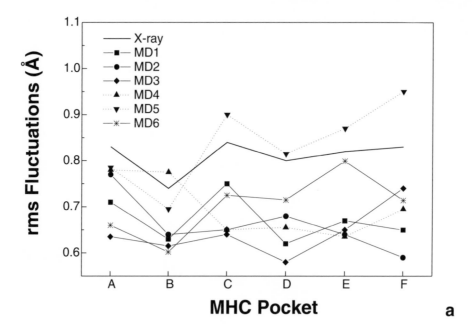

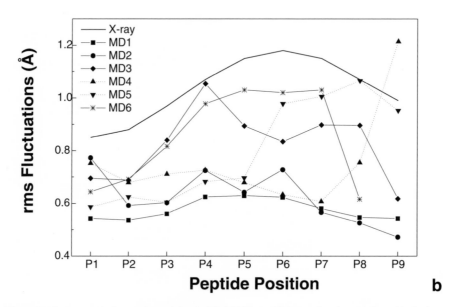

Figure 1. a) rms atomic fluctuations of the six HLA-B27 specificity pockets in complex with the bacterial peptides. The values for the X-ray structure (bold line) were obtained from temperature factors. b) rms atomic fluctuations of HLA-B27-bound peptides (backbone atoms) for the crystal structures and six MD conformations.

from the peptide to the binding sites. The situation is much more dramatic for specificity pocket F, which normally binds residue 9. There is a difference of nearly 0.4 Å in atomic fluctuations, which indicates a large movement of the pocket and hence no interactions to the peptide's C-terminal residue. In summary, it can be seen that the highest fluctuation values are found for the complex with inactive peptides.

The lower graph in Fig. 1 provides a complementary picture, showing the fluctuation of the binding ligands, the peptides. The graph is even easier to interpret; low atomic fluctuations indicate tight binding, and vice versa. Again, the most active peptides 1, 2, and 3, show the lowest fluctuations.

Much more interesting in this context is the behavior of the octapeptide. This is also an active peptide, and thus needs tight binding to the MHC molecule in order to be presented to the T-cell receptor. The octapeptide (*) shows a highly fluctuating sequence in the middle, namely for the residues 4 to 7. Amino acids at position 1, 2, and 8, however, are at very low fluctuation levels.

Thus, the octapeptide reveals the importance of the binding pockets for peptide presentation, the binding to pockets A, B, and F being complementary to residues 1 and 2, while the C-terminus of the ligands seems the precondition for presentation to T-cell receptor.

5.4.3 Solvent-Accessible Surface Areas

For peptide side chains binding to a pocket within the surface of a receptor, the idea of correlating this process with the residual surface that is accessible to the solvent, seems straightforward. As we learned from the X-ray studies [7], binding pockets B and F that bind residues 2 and the C-terminus 1 of the peptidic ligands really bury these side chains. Thus, most of the side chain-solvating molecules must be removed and replaced by an interaction to the side chains of the receptor protein that make up the walls of the binding pocket. The residual surface that is still accessible to the solvent after the ligand has been docked, is a measure of the depth of binding into the pocket. It correlates with the tightness and more or less with the strengths and number of binding interactions with the pocket.

Accessible and buried surface areas, respectively, were computed using an algorithm from Connolly [11]. A probe atom with a 1.4 Å radius is used to walk around the ligand or the part that is visible. The radius of the probes simulates a water molecule. In order to quantify the results in terms of percentages, free peptides of the sequence Gly-Xaa-Gly were built in an extended conformation and computations performed with those in a similar fashion. These served as an example of fully solvated or fully accessible reference surfaces.

Using this description technique, the important binding pockets of the MHC molecules could be easily identified by analyzing the ligand–protein complexes. Fig. 2 shows the results of the study on B*2705–ligand interaction. Again, the active peptides are represented by solid lines, the inactive ones by open lines; the octapeptide is represented by *. The horizontal line in the graph represents a 50%-buried residue. It is immediately clear from the graph that residues at position 1, 2, and 9 are the important binding locations of the MHC molecule, because they are buried almost completely. Significantly this is not true for the low-affinity peptides, since their side chains, even at the C-terminus, are 70% exposed to the solvent. Again,

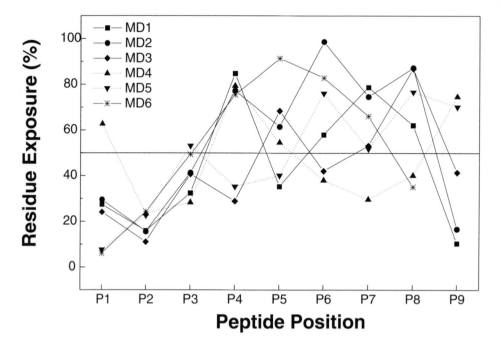

Figure 2. Solvent-accessible surface area of MHC-bound nonapeptides. All values were computed for relaxed molecular dynamics mean structures, time-averaged over 500 conformations during the last 50 ps of the molecular dynamics simulation.

the graph provides an immediate interpretation of the pharmacophore of the peptidic ligands. Residues that bind to MHC can be identified by their high degree of burying; inactive peptides can be seen not to be buried at all at these positions. The side chains of the ligands that bind to the T-cell receptor must not be buried into the MHC-binding cleft; hence they can be seen exposed to the solvent. This is true for the residues 4–7 of the active, high-affinity peptides.

Thus, the solvent-accessible surface correlates closely with the experimental observations and seems to be an excellent tool for careful and detailed interpretation of ligand–protein interaction.

5.4.4 Interaction Energies

The analysis of interaction energies has been broadly discussed and is still debated for instance in respect of hydrogen bond energies. Many authors consider that the calculation of interaction energies is useless because of the weakness of the potentials used for the calculation. The simple Coulomb equation is mostly taken for the electrostatics, hydrophobic interaction is neglected in most cases. Nevertheless, some calculations, if comparing several structures, at least provide an estimation, whether an energetically favored interaction occurs or not.

However, one should always bear in mind, that the computation of interaction energies sums up very large numbers to result tiny differences. This means, the calculation may be extremely sensitive to the choice of starting conditions, like geometries, or the quality of charge calculations.

In the present case, we have studied the energies of the protein–ligand interaction complex in terms of energy value per residue for charged residues, neutral residues, the whole peptide, and for the water molecules. The internal energy of each of these substructures has been computed every 20 ps throughout the molecular dynamics simulation.

The idea was that artefacts caused by strong interactions should be recognized by an increase of internal energy for the neutral amino acids, because those would have to compensate the structure artefacts. The analyses showed that all substructures had decreased their internal energies.

Most interesting was the finding that the energy of the bound peptides fluctuates quite widely but in general falls to a level that is more than 30% lower in energy than the starting structure which had been energy-minimized before docking. Again, this shows the performance of ligand optimization by molecular dynamics simulation in the presence of restraints from the binding site.

The complete situation has always been analyzed energetically. The vacuum, or solvent energy of the ligands may be very different from the minimum energy of the docked ligand. The docked conformation, however, is the only valid optimized structure which can be used for further drug design steps. The solvent conformations, and especially the vacuum conformations, are in most cases absolutely useless. There are of course exceptions, for instance for rigid molecules!

Also remarkable was the overall decrease in internal energy for the charged residues, which was computed to be about 60 kJ mol^{-1} per residue. The interactions of the bound peptide to neutral residues, or to itself, fluctuated and changed only slightly. This seems reasonable because during the docking process hydrophobic interactions were carefully optimized knowing the weakness of the potentials. Thus, no dramatic changes are to be expected during the molecular dynamics simulation of the interaction complex.

Although interaction energies might not be able to provide a quantification of the protein–ligand interactions—as had been previously hoped—we feel that they allow for an estimation of quality and provide a feeling for what happened during the molecular dynamics simulation.

There remains an open question whether more sophisticated calculations of interaction energies—for instance using "good" charge calculation that take into account the Poisson–Boltzmann equation—are able to improve the quality of interaction energy calculations. This is also true for quantum chemical interaction energy calculations. Although there are many counter-arguments, especially with respect to lacking performance of an ab initio basis set for non-covalent interactions, good results have been obtained even with very simple basis sets using quantum mechanical calculations. One such example is the prediction of ammonium partial structure interaction with an aromatic moiety by semiempirical methods as early as 1975 [12]. This interaction has been fully confirmed by the X-ray structure of acetylcholine esterase some 20 years later [13].

5.5 SAR of the Antigenic Peptides from Molecular Dynamics Simulations and Design of Non-natural Peptides as High-Affinity Ligands for a MHC I Protein

From the analyses described in the previous section, much information was provided for a design of non-natural ligands. Knowledge of the site, flexibility and side chain interaction of the binding pockets led to the idea to investigate whether all of them were used optimally by the native ligands. It evolved that the binding pocket D, responsible for interacting with side chains in position 3 of the peptidic ligands, provides much more space than the native ligands used to fill. This might provide the possibility of adding binding interactions by placing a larger side chain or substituent in this pocket. This would increase binding energy and hence lead to a ligand with high affinity.

Pocket D is hydrophobic in nature, lined by tyrosine, histidine, leucine, tyrosine and leucine. As seen in the crystal structure-based homology models, the ligand's side chains only interact with the upper rim of the hydrophobic pocket, taking into account only both tyrosines.

5.5.1 The Design of New Ligands

To rationalize the binding of a hydrophobic side chain to pocket D, we have computed the optimum interaction site. The computation was made for the isolated pocket using the program GRID (see section 4.6.2) and applying the methyl group as a probe. The resolution of the grid was 0.5 Å. Interaction of the methyl probe with the pocket walls was summed to result in a contour plot localizing negative binding energy. This indicates the approximate size of a putative ligand to fill the pocket completely and to interact with the pocket by gain of interaction energy (Fig. 1).

As we had hoped, the contour map extends much more deeply into the pocket, than the native ligand's side chain. Thus, it could accomodate even larger residues than phenylalanine for instance. Using molecular graphics, we were able to show that residues as bulky as naphthylalanine fitted nicely into the pocket. Furthermore, we predicted that this should be possible without distortion of the geometry of the peptide ligand's backbone.

The question whether the hypothezised interactions remain stable—as suggested by this static picture—was addressed by molecular dynamics of the solvated complexes. One of the bacterial peptides (Lys-Arg-Gly-Ile-Asp-Lys-Ala-Ala-Lys) was used as a template and position 3 substituted by apolar side chains of increasing size (Table 1).

Analysis techniques, described earlier, were then applied to determine, whether the constructs would be expected to be stable and worth synthesizing. Molecular dynamics simulation in water for 150 ps revealed for all cases that the new substituents in pocket D did not affect the 3D structure of the binding groove—at least, not to an extent that would destroy the complex. rms deviations from X-ray structure and from previous homology models, respectively, were reasonably low.

The analysis of the surface areas indicated the stability of the complexes with the new, non-natural side chains in position 3 (Fig. 2(a)). Buried surfaces (the opposite of solvent-accessible surface) of more than 100 $Å^2$ are found for the anchor positions 1, 2, and 9 corresponding to

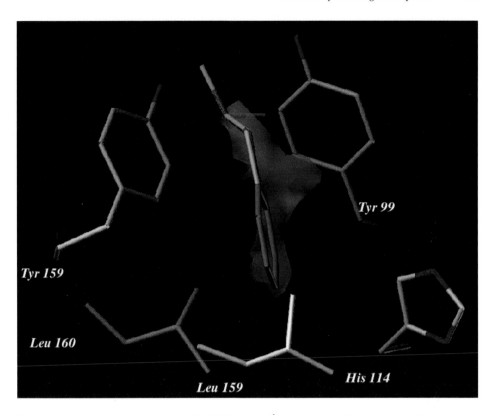

Figure 1. Energy contours indicating at the 2.75 kcal mol⁻¹ level the most favorable interactions between the free pocket D and a methyl probe. Complementary peptide side chains were fitted *a posteriori* to the energy contour map from the crystal structure of HLA-B27-bound nonapeptide (isoleucine, blue; homophenylalanine, red).

Table 1. Sequence of non-natural HLA-B27 ligands derived from two bacterial nonapeptides[c]

Peptide no.	Peptide sequence	Symbol
	Lys-Arg————**Xaa**————Ile-Asp-Lys-Ala-Ala-Lys	
1	Gly[a]	▲
2	Leu	▼
3	Hpa	◆
4	Ana	●
5	Bna	■
	Gln-Arg-Leu ——— **Spacer** ——— Lys	
6	Lys-Glu-Ala-Ala-Glu[b]	
7	Gly-Gly-Gly-Gly-Gly	
8	Aba-Aba-Aba	
9	Aha-Aha	

[a] *Chlamydia trachomatis* groEl 117-125 [14].
[b] *Escherichia coli* dnaK 220-228.
[c] Hpa = Homophenylalanine; Ana = α-Naphthylalanine; Bna = β-Naphthylalanine;
Aba = 4-Aminobutyrie acid; Aha = 6-Aminohexanoic acid.

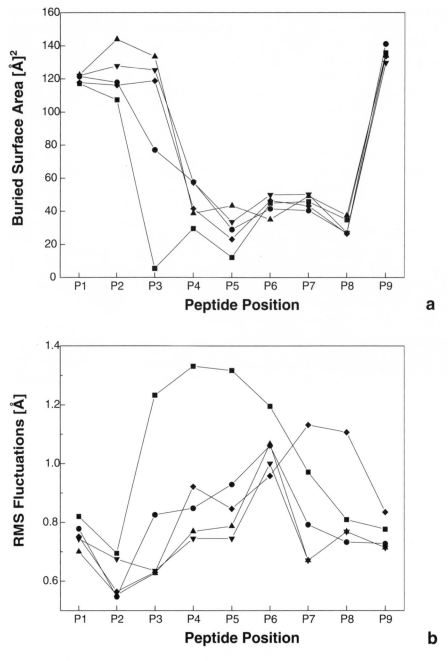

Figure 2. Atomic properties of HLA-B27-bound nonapeptides 1–5 simulated by 150 ps molecular dynamics simulation on solvated MHC–peptide pairs. a) Buried surface areas of HLA-B27-bound peptides, calculated from relaxed time-averaged conformations. b) Rms atomic fluctuations averaged per peptide residue over all atoms and calculated from time-averaged conformations. Symbols used explained in Table 1.

the binding pockets at the N- and C-termini. As expected the middle part of the ligand (residues 4–7) shows low buried surface areas. Insofar, the picture closely resembles the normal situation. A striking difference occurs at position 3 of the ligand, corresponding to pocket D. Correlated with the size of the residues the buried surface increases significantly up to 140 Å^2. By comparison the native glycine at position 3 is less than 10 Å^2. Thus, the analysis of the surface areas directly reflects the GRID calculations and the expectations arising from the calculations, that the non-natural side chains, as do the naphthyl derivatives, should be able to fill and stabilize binding at pocket D better than the native residue.

A very similar situation results from the analysis of the atomic fluctuations. As expected lowest fluctuations are found for anchor positions 1, 2, and 9. High mobility occurs at positions 4–7, representing the part of the peptide ligand pointing to the solvent or the T-cell receptor, respectively. At position 3, however, a clear discrimination is again possible between the native and the new, synthetic ligands. Atomic fluctuations at position 3 are strictly related to the size of the residue and hence to the number and strength of interactions to pocket D (Fig. 2(b)). Again, the most flexible ligand is the parent peptide having a glycine at position 3. In contrast, naphthylalanine residues restrain the whole peptide in its mobility by their tight binding to pocket D. Therefore, our prediction was to expect higher affinity for ligands bearing non-natural side chains in position 3. Those side chains should have aromatic/hydrophobic properties for optimal interaction with the residues that comprise the walls of the binding pocket in the MHC molecule.

5.5.2 Experimental Validation of the Designed Ligands

In order to prove the model, all five isomers have been synthesized and purified [14]. The binding assay was based on an immuno reaction. Antibodies raised against the MHC–protein–ligand complex are able to recognize the geometry of polypeptide chain that makes up the active site. As it has been shown experimentally (see earlier), MHC molecules fold up in the presence of a ligand. Therefore, non-binding ligands should lead to a significantly reduced number of correctly folded active sites to be recognized by the antibodies. Those conformation-specific antibodies could be produced and shown to be effective in the present case [14, 15]. In this assay the non-natural peptides showed a significant increase in binding, reaching the levels of native high-affinity binders like GROE 1 from bacteria [14].

So far, the experimental data fully supported the hypothesis that had been derived from theoretical studies. It should be noted, however, that this success is only a small part of the whole story. Detailed inspection of the experimental results reveals that glycine and leucine at position 3 have equivalent binding affinities. They were predicted, however, to interact differently with pocket D. The reason for the underestimation of the glycine interaction are several. On one hand, the model does not represent the complete situation of T-cell receptor–MHC interaction; on the other, entropic contributions are omitted from the force field calculations. They may nevertheless contribute significantly to the binding of the residues to MHC. The higher conformational flexibility may in this case compensate entropically for the lack of van der Waals interactions.

5.6 Summary and Conclusion

The description of the properties of the ligand–protein complex was the central point in this chapter. By carefully applying theoretical tools such as solvent-accessible surface calculation or consideration of atomic fluctuation, a detailed picture of the protein–ligand interaction could be established. It is important to note, that the static inspection of a model that emerged from homology modeling is not sufficient to derive a conclusion for the design of new ligands. Molecular dynamics simulation of the complexes of several ligands, in only a fully solvated environment, identified the important anchor positions or parts of the molecule that could be neglected for binding optimizations.

In our view this reflects a general problem. Very often only the crystal structure is taken as sufficient for the design of new ligands. In contrast, we found it very useful to carry out extensive and time-consuming molecular dynamics simulations in a fully solvated state, as only these calculations revealed the further binding details that ultimately were important for the design. Moreover, only the molecular dynamics simulations provided tools as atomic mobility studies, that allowed an estimation of the binding abilities of the new ligands to be made. Interaction energies were found instead to be much less significant for judging the designed ligands.

It should be pointed out that the experiments supported the hypotheses derived from the models and that the whole study was completely prospective, both in the mechanistic interpretation and the design step. From our viewpoint this shows the power of structure-based design and its ability to produce optimized or even entirely new ligands. Our recent studies show that protein–ligand complexes, established by homology modeling can be used successfully to introduce non-peptidic parts into peptide ligands, or even to create heterocyclic non-peptidic ligands using de novo design procedures.

References

[1] Nicklaus, M. C., Wang, S., Driscoll, J. S., and Milne, G. W. A. *Bioorganic Med. Chem.* **3**, 411–428 (1995).
[2] Rognan, D., Reddehase, M. J., Koszinowski, U. H., and Folkers, G. *Proteins* **13**, 70–85 (1992).
[3] Sufrin, J.,R., Dunn, D. A., and Marshall, G. R. *Mol. Pharmacol.* **19**, 307–313 (1981).
[4] Lambert, M. H., and Scheraga, H. A. *J. Comput. Chem.* **10**, 770–797 (1989).
[5] Fauchère, J. L., and Pliska, V. *Eur. J. Med. Chem.* **18**, 369–375 (1983).
[6] Rognan, D., Zimmermann, N., Jung, G., and Folkers, G. *Eur. J. Biochem.* **208**, 101–113 (1992).
[7] Madden, D. R., Gorga, J. C., Strominger, J. L., and Wiley, D.C. *Cell* **70**, 1035–1048 (1992).
[8] Rognan, D., Scappoza, L., Folkers, G., and Daser, A. *Biochemistry* **33**, 11476–11485 (1994).
[9] Rognan, D., Scappoza, L., Folkers, G., and Daser, A. *Proc. Natl. Acad. Sci. U.S.A.* **92** , 753–757 (1995).
[10] Honig, B., and Nicholls, A. *Science* **268**, 1144–1149 (1995).
[11] Connolly, M. L. *J. Appl. Crystallogr.* **16**, 548–558 (QCPE Programme No. 429).
[12] Höltje, H.-D., and Kier, L. B. *J. Pharm. Sci.* **64**, 418–423 (1975).
[13] Sussman, J. L., Harel, M., Frolow, F., Oefner, C., Goldman, A., Toker, L., and Silman, I. *Science* **253**, 872–875 (1991).
[14] Daser, A., Henning, U., and Heuklein, P. *Mol. Immunol.* **31**, 331–336 (1994).
[15] Ellis, S. A., Taylor, C., and McMichael, A. J. *Human Immunol.* **5**, 49–59 (1982)   .

Appendix

Appendix 1: Program IXGROS

```c
/* ixgros 2.0*/
/* (c) 1993 W. Sippl */

#include <stdio.h>
#include <stdlib.h>

#define TRUE 1
#define FALSE 0

#ifndef EXIT_FAILURE
#define EXIT_FAILURE -1
#endif

#define LOADMODE 0
#define SAVEMODE 1

typedef int boolean;

typedef struct {
    FILE *file;
    int nbonds,nrings;
    short flag;
    short *oldtuple;
    boolean energies,first_data,first_rbond,first_ring;
} AFILE;

typedef struct {
    int     tuple[15];
    float     energy;
} cnfrow;
```

```
typedef struct {
    int    a1,a2,a3,a4;
    int    incr,rbid;
    float refangle;
} rotbond;
```

/* PROTOS */

```
AFILE* SRCH_AFILE_OPEN(char*,int);
boolean SRCH_AFILE_CLOSE (AFILE*);
boolean SRCH_AFILE_READ_HEADER(AFILE*, boolean*, int*,int*,int*, char*);
boolean SRCH_AFILE_WRITE_HEADER(AFILE*, boolean, int,int,int, char*);
boolean SRCH_AFILE_READ_DATA(AFILE*,int*,float*);
boolean SRCH_AFILE_WRITE_DATA(AFILE*,int*,double);
boolean SRCH_AFILE_READ_RBOND(AFILE*,int*,int*,int*,int*,int*,int*,float*);
boolean SRCH_AFILE_WRITE_RBOND(AFILE*,int*,int*,int*,int*,int*,int*,float*);
```

```
boolean TestRange(int,int);
void FindNeighbours();
```

/* END OF PROTOS */

/*declaration of variables*/

```
AFILE *infile;         /* .ANG-input file   */
AFILE *outfile;        /* .ANG-output file  */
FILE *asciiout;        /* ASCI file         */
```

```
boolean has_energy;    /* energies              */
int numbonds,          /* number of rotable bonds     */
    numrings,          /* number of rings   */
    numconfs;          /* number of conformations    */
```

```
int mincnt=0;          /* number of minima */
```

```
char molname[20];      /* molecule name     */
```

```
rotbond RB[10];        /* arrays for rotable bonds */
```

```
cnfrow *conformations;
int i,count;
```

```
/********************************************************
* find neighbours *
********************************************************/
boolean TestRange(int ref, int num)
{
    boolean isin=FALSE;
    int    testwert=0,
           localref1=0,
    l      ocalref2=0,
           cnt=0;

        for (cnt=0;cnt<numbonds;cnt++)
        {

        testwert = conformations[num].tuple[cnt];

        localref1 = conformations[ref].tuple[cnt] + RB[cnt].incr;
        localref2 = conformations[ref].tuple[cnt] – RB[cnt].incr;

        isin = (testwert <= localref1) && (testwert >= localref2);

                if (isin==FALSE)
            {
            if (localref1 >= 360)
            {
            testwert=testwert+360;
            }
            if (localref2 <= 0)
            {
            testwert=testwert-360;
            }

            isin = (testwert <= localref1) && (testwert >= localref2);
            if (isin==FALSE)
            {
            return FALSE;
            }
        }
    }
    return isin;
}
```

```
void FindNeighbours()
{
    signed int i,TestKandidat,Referenz;
    boolean minfound=TRUE;
    double tmpreal=0.0;

    for(Referenz=0;Referenz<numconfs;Referenz++) {
        minfound = TRUE;

            for(TestKandidat=0;minfound && (TestKandidat<numconfs);TestKandidat++) {

        if (TestRange(Referenz,TestKandidat)==TRUE) {
            if (conformations[TestKandidat].energy < conformations[Referenz].energy) {
                minfound = FALSE;
            }
        }
    }

    if (minfound==TRUE) {
        tmpreal = conformations[Referenz].energy;
        SRCH_AFILE_WRITE_DATA(outfile,&conformations[Referenz].tuple[0],tmpreal);
        mincnt=mincnt+1;

        fprintf(asciiout,"%-5d",mincnt);

        fprintf(asciiout,"KNum: %-6d ,,Referenz+1);

        for(i=0;i<numbonds;i++) {
        fprintf(asciiout," %3d ,,conformations[Referenz].tuple[i]);
        }

        fprintf(asciiout," %f\n",conformations[Referenz].energy);
        }
    }
}
/* main program */
int main(int argc, char *argv[])
{
    int i;
    char anginname[50],
    angoutname[50],
    ascoutname[50];

    if (argc != 2)
```

```
{
    printf(„Usage: %s [filename]\n“,argv[0]);
    puts(„Copyright (c) 1993 W. Sippl“);
    exit(EXIT_FAILURE);
}

strcpy(anginname,argv[1]);
strcpy(angoutname,argv[1]);
strcpy(ascoutname,argv[1]);
asciiout = fopen(strcat(ascoutname,"_fam.asc“),"w“);
if (asciiout==NULL)
{
    printf(„Cannot open output file %s .\n“,strcat(argv[1],"_fam.asc“));
    exit(EXIT_FAILURE);
}

fprintf(asciiout,"\nixgros 2.0\nCopyright (c) 1993 W. Sippl\n\n“);
fprintf(asciiout,"Filename: %s\n“,argv[1]);

infile = SRCH_AFILE_OPEN(strcat(anginname,".ang“),LOADMODE);
if (infile==NULL)
{
    printf(„Cannot open input file %s.\n“,strcat(argv[1],".ang“));
    exit(EXIT_FAILURE);
}

outfile = SRCH_AFILE_OPEN(strcat(angoutname,"_fam.ang“),SAVEMODE);
if (outfile==NULL)
{
    printf(„Cannot open output file %s.\n“,strcat(argv[1],"_fam.ang“));
    exit(EXIT_FAILURE);
}
    if
(SRCH_AFILE_READ_HEADER(infile,&has_energy,&numbonds,&numrings,&numconfs,
molname))
    {
        fprintf(asciiout,"numbonds = %d, numrings = %d, numconfs = %d\n“,
            numbonds,numrings,numconfs);
        fprintf(asciiout,"molecule name: %s\n“,molname);
    }
    else
    {
        puts(„Reading not possible – exit“);
        exit(EXIT_FAILURE);
    }
```

```
conformations = malloc(numconfs * sizeof(cnfrow));

for (i=0;i<numbonds;i++)
{
    SRCH_AFILE_READ_RBOND(infile,&RB[i].a1,&RB[i].a2,&RB[i].a3,&RB[i].a4,
        &RB[i].incr,&RB[i].rbid,&RB[i].refangle);

    fprintf(asciiout,"A1-4 % 4d % 4d % 4d % 4d, Incr = % 4d, RefAngle = % f\n",
        RB[i].a1,RB[i].a2,RB[i].a3,RB[i].a4,RB[i].incr,RB[i].refangle);
}

/* reading conformations */
for (i=0;i<numconfs;i++)
{
    SRCH_AFILE_READ_DATA(infile,&conformations[i].tuple[0],&conformations[i].energy);
}
fprintf(asciiout,"\nminima\n\n");
if ( ! SRCH_AFILE_WRITE_HEADER(outfile,has_energy,numbonds,0,mincnt,molname)) {
    printf("Cannot write header.\n");
}


FindNeighbours();

if ( ! SRCH_AFILE_WRITE_HEADER(outfile,has_energy,numbonds,0,mincnt,molname)) {
    printf("Cannot write header.\n");
}

if (infile!=NULL)
{
    SRCH_AFILE_CLOSE(infile);
}

if (outfile!=NULL)
{
    for(i=0;i<numbonds;i++) {
        SRCH_AFILE_WRITE_RBOND(outfile,RB[i].a1,RB[i].a2,RB[i].a3,
            RB[i].a4,RB[i].incr,RB[i].rbid,RB[i].refangle);
    }

    SRCH_AFILE_CLOSE(outfile);

}
```

```
/* close ASCI file */
if (asciiout!=NULL)
{
fclose(asciiout);
}


if (free!=NULL)
{
free(conformations);
}


}
```

Appendix 2: Brookhaven data file of trypsin complexed with a benzamidine inhibitor.

For more clarity some ATOMS (no. 11-1654) and HETATMS (no. 1680-1824) have been omitted.

```
REMARK   1 REFERENCE 2                                                    2TBS   23
REMARK   1   AUTH   A.O.SMALAS,A.HORDVIK,L.K.HANSEN,E.HOUGH,K.JYNGE        2TBS   24
REMARK   1   TITL   CRYSTALLIZATION AND PRELIMINARY X-RAY                  2TBS   25
REMARK   1   TITL 2 CRYSTALLOGRAPHIC STUDIES OF BENZAMIDINE-INHIBITED      2TBS   26
REMARK   1   TITL 3 TRYPSIN FROM THE NORTH ATLANTIC SALMON (SALMO          2TBS   27
REMARK   1   TITL 4 SALAR)                                                 2TBS   28
REMARK   1   REF    J.MOL.BIOL.                V. 214    355 1990          2TBS   29
REMARK   1   REFN   ASTM JMOBAK   UK ISSN 0022-2836                 0?0    2TBS   30
REMARK   1 REFERENCE 3                                                     2TBS   31
REMARK   1   AUTH   M.MARQUART,J.WALTER,J.DEISENHOFER,W.BODE,R.HUBER       2TBS   32
REMARK   1   TITL   THE GEOMETRY OF THE REACTIVE SITE AND OF THE           2TBS   33
REMARK   1   TITL 2 PEPTIDE GROUPS IN TRYPSIN, TRYPSINOGEN AND ITS         2TBS   34
REMARK   1   TITL 3 COMPLEXES WITH INHIBITORS                             2TBS   35
REMARK   1   REF    ACTA CRYSTALLOGR.,SECT.B      V.  39    480 1983       2TBS   36
REMARK   1   REFN   ASTM ASBSDK   DK ISSN 0108-7681                 62￼    2TBS   37
REMARK   2                                                                 2TBS   38
REMARK   2 RESOLUTION. 1.8  ANGSTROMS.                                     2TBS   39
REMARK   3                                                                 2TBS   40
REMARK   3 REFINEMENT.                                                     2TBS   41
REMARK   3   PROGRAM                      PROLSQ                           2TBS   42
REMARK   3   AUTHORS                      KONNERT,HENDRICKSON              2TBS   43
REMARK   3   RMSD BOND DISTANCES          0.020   ANGSTROMS                2TBS   44
REMARK   3   RMSD BOND ANGLE DISTANCES    0.040   ANGSTROMS                2TBS   45
REMARK   3                                                                 2TBS   46
REMARK   3   NUMBER OF REFLECTIONS        14474                            2TBS   47
REMARK   3   RESOLUTION RANGE        6.0 - 1.8   ANGSTROMS                 2TBS   48
REMARK   3   DATA CUTOFF                  3.0     SIGMA(F)                 2￼BS   49
REMARK   3                                                                 2TBS   50
REMARK   3   NUMBER OF PROTEIN ATOMS                          1622         2TBS   51
REMARK   3   NUMBER OF SOLVENT ATOMS                           164         2TBS   52
REMARK   3                                                                 2TBS   53
REMARK   3   RMS DEVIATIONS FROM IDEAL VALUES (THE VALUES OF               2TBS   54
REMARK   3      SIGMA, IN PARENTHESES, ARE THE INPUT ESTIMATED             2TBS   55
REMARK   3      STANDARD DEVIATIONS THAT DETERMINE THE RELATIVE            2TBS   56
REMARK   3      WEIGHTS OF THE CORRESPONDING RESTRAINTS)                   2TBS   57
REMARK   3   DISTANCE RESTRAINTS (ANGSTROMS)                               2TBS   58
REMARK   3      BOND DISTANCE                          0.020(0.01)         2TBS   59
REMARK   3      ANGLE DISTANCE                         0.040(0.02)         2TBS   60
REMARK   3      PLANAR 1-4 DISTANCE                    0.046(0.03)         2TBS   61
REMARK   3   PLANE RESTRAINT (ANGSTROMS)               0.015(0.02)         2TBS   62
REMARK   3   CHIRAL-CENTER RESTRAINT (ANGSTROMS**3)    0.042(0.12)         2TBS   63
REMARK   3   NON-BONDED CONTACT RESTRAINTS (ANGSTROMS)                     2TBS   64
REMARK   3      SINGLE TORSION CONTACT                 0.172(0.50)         2TBS   65
REMARK   3      MULTIPLE TORSION CONTACT               0.321(0.50)         2TBS   66
REMARK   3      POSSIBLE HYDROGEN BOND                 0.221(0.50)         2TBS   67
REMARK   3   CONFORMATIONAL TORSION ANGLE RESTRAINT (DEGREES)              2TBS   68
REMARK   3      PLANAR                                4.(  3.0)            2TBS   69
REMARK   3      STAGGERED                             15.2(15.0)           2TBS   70
REMARK   3      ORTHONORMAL                           27.6(20.0)           2TBS   71
REMARK   4                                                                 2TBS   72
REMARK   4 THE STRUCTURE WAS SOLVED AND REFINED WITH ONLY A SMALL          2TBS   73
REMARK   4 FRACTION OF THE PRIMARY STRUCTURE KNOWN.  HOWEVER, THE GENE     2TBS   74
REMARK   4 SEQUENCE OF SALMON TRYPSIN HAS NOW BECOME AVAILABLE AND         2TBS   75
REMARK   4 SOME DISCREPANCIES BETWEEN THIS SEQUENCE AND SEQUENCE           2TBS   76
REMARK   4 OBTAINED FROM THE X-RAY STUDIES INDICATE THAT THE               2TBS   77
REMARK   4 MENTIONED SEQUENCES MAY CORRESPOND TO ISO-ENZYMES.             2TBS   78
REMARK   5                                                                 2TBS   79
REMARK   5 THE STRUCTURE WAS SOLVED BY MOLECULAR REPLACEMENT METHODS,      2TBS   80
```

```
REMARK   5 USING THE MERLOT-PACKAGE (FITZGERALD, P.  (1988) J.APPL.      2TBS  81
REMARK   5 CRYST., 21, 273-278), AND THE REFINED MODEL OF               2TBS  82
REMARK   5 BOVINE TRYPSIN AS SEARCH MODEL (PROTEIN DATA BANK, ENTRY      2TBS  83
REMARK   5 3PTB).                                                        2TBS  84
REMARK   6                                                              2TBS  85
REMARK   6 THE CALCIUM IS ION IS BOUND IN A MANNER SIMILAR TO THAT       2TBS  86
REMARK   6 OBSERVED IN BOVINE TRYPSIN.                                   2TBS  87
REMARK   7                                                              2TBS  88
REMARK   7 THE AMINO ACID NUMBERING SCHEME USED IS ADOPTED FROM          2TBS  89
REMARK   7 CHYMOTRYPSINOGEN.                                             2TBS  90
REMARK   8                                                              2TBS  91
REMARK   8 THIS ENTRY WAS REFINED USING A NON-STANDARD SETTING FOR       2TBS  92
REMARK   8 THE SPACE GROUP P 21 21 2.  THE FOLLOWING SYMMETRY            2TBS  93
REMARK   8 OPERATORS MUST BE USED TO GENERATE CRYSTALLOGRAPHICALLY       2TBS  94
REMARK   8 RELATED MOLECULES.                                            2TBS  95
REMARK   8             X,      Y,     Z                                  2TBS  96
REMARK   8            -X,  1/2+Y,    -Z                                  2TBS  97
REMARK   8         1/2+X,     -Y,    -Z                                  2TBS  98
REMARK   8         1/2-X,  1/2-Y,     Z                                  2TBS  99
REMARK   9                                                              2TBS 100
REMARK   9 THERE IS NO SIDE-CHAIN DENSITY BEYOND CB FOR LYS 23,          2TBS 101
REMARK   9 ARG 62, LYS 74, TYR 97, ASN 178.                             2TBS 102
REMARK  10                                                              2TBS 103
REMARK  10 THERE IS NO CLEAR DENSITY FOR THE LAST THREE C-TERMINAL       2TBS 104
REMARK  10 RESIDUES ALA 243, SER 244, TYR 245.                          2TBS 105
REMARK  11                                                              2TBS 106
REMARK  11 THE PRIMARY SEQUENCE OF THIS FORM OF SALMON TRYPSIN IS        2TBS 107
REMARK  11 DEPOSITED IN THE EMBL DATA LIBRARY: ACCESSION NO. X70071.     2TBS 108
REMARK  12                                                              2TBS 109
REMARK  12 PDB ADVISORY NOTICE:                                         2TBS 110
REMARK  12 OH TYR 245 AND N ALA 28 HAVE VERY SHORT CONTACTS WITH         2TBS 111
REMARK  12 WATER MOLECULES 388 AND 418 RESPECTIVELY.                     2TBS 112
SEQRES   1    222  ILE VAL GLY GLY TYR GLU CYS LYS ALA TYR SER GLN ALA   2TBS 113
SEQRES   2    222  HIS GLN VAL SER LEU ASN SER GLY TYR HIS PHE CYS GLY   2TBS 114
SEQRES   3    222  GLY SER LEU ASN GLU ASN TRP VAL VAL SER ALA ALA       2TBS 115
SEQRES   4    222  HIS CYS TYR LYS SER ARG VAL GLU VAL ARG LEU GLY GLU   2TBS 116
SEQRES   5    222  HIS ASN ILE LYS VAL THR GLU GLY SER GLU GLN PHE ILE   2TBS 117
SEQRES   6    222  SER SER SER ARG VAL ILE ARG HIS PRO ASN TYR SER SER   2TBS 118
SEQRES   7    222  TYR ASN ILE ASP ASN ASP ILE MET LEU ILE LYS LEU SER   2TBS 119
SEQRES   8    222  LYS PRO ALA THR LEU ASN THR TYR VAL GLN PRO VAL ALA   2TBS 120
SEQRES   9    222  LEU PRO THR SER CYS ALA PRO ALA GLY THR MET CYS THR   2TBS 121
SEQRES  10    222  VAL SER GLY TRP GLY ASN THR MET SER SER THR ALA ASP   2TBS 122
SEQRES  11    222  SER ASP LYS LEU GLN CYS LEU ASN ILE PRO ILE LEU SER   2TBS 123
SEQRES  12    222  TYR SER ASP CYS ASN ASP SER TYR PRO GLY MET ILE THR   2TBS 124
SEQRES  13    222  ASN ALA MET PHE CYS ALA GLY TYR LEU GLU GLY GLY LYS   2TBS 125
SEQRES  14    222  ASP SER CYS GLN GLY ASP SER GLY GLY PRO VAL VAL CYS   2TBS 126
SEQRES  15    222  ASN GLY GLU LEU GLN GLY VAL VAL SER TRP GLY TYR GLY   2TBS 127
SEQRES  16    222  CYS ALA GLU PRO GLY ASN PRO GLY VAL TYR ALA LYS VAL   2TBS 128
SEQRES  17    222  CYS ILE PHE SER ASP TRP LEU THR SER THR MET ALA SER   2TBS 129
SEQRES  18    222  TYR                                                   2TBS 130
HET    BEN  246        9        BENZAMIDINE INHIBITOR                    2TBS 131
HET     CA  247        1     CALCIUM +2 COUNTER ION                      2TBS 132
FORMUL   2  BEN    C7 H8 N2                                              2TBS 133
FORMUL   3   CA    CA1                                                   2TBS 134
FORMUL   4  HOH    *164(H2 O1)                                           2TBS 135
SSBOND   1 CYS     22    CYS    157                                      2TBS 136
SSBOND   2 CYS     42    CYS     58                                      2TBS 137
SSBOND   3 CYS    128    CYS    232                                      2TBS 138
SSBOND   4 CYS    136    CYS    201                                      2TBS 139
SSBOND   5 CYS    168    CYS    182                                      2TBS 140
SSBOND   6 CYS    191    CYS    220                                      2TBS 141
CRYST1   61.950   84.330   39.110  90.00   90.00  90.00 P 21 21 2    4   2TBS 142
ORIGX1      1.000000  0.000000  0.000000        0.00000                 2TBS 143
ORIGX2      0.000000  1.000000  0.000000        0.00000                 2TBS 144
ORIGX3      0.000000  0.000000  1.000000        0.00000                 2TBS 145
SCALE1      0.016142  0.000000  0.000000        0.00000                 2TBS 146
SCALE2      0.000000  0.011858  0.000000        0.00000                 2TBS 147
```

```
SCALE3          0.000000   0.000000   0.025569          0.00000                      2TBS 148
ATOM      1  N    ILE   16   29.847  42.460  26.572  1.00  8.08              2TBS 149
ATOM      2  CA   ILE   16   29.126  43.406  27.446  1.00 13.82              2TBS 150
ATOM      3  C    ILE   16   29.929  43.832  28.670  1.00 17.40              2TBS 151
ATOM      4  O    ILE   16   30.273  43.035  29.550  1.00  9.27              2TBS 152
ATOM      5  CB   ILE   16   27.690  42.882  27.913  1.00 10.91              2TBS 153
ATOM      6  CG1  ILE   16   26.845  42.492  26.687  1.00  8.97              2TBS 154
ATOM      7  CG2  ILE   16   26.956  43.882  28.864  1.00  9.09              2TBS 155
ATOM      8  CD1  ILE   16   26.415  43.592  25.702  1.00 13.04              2TBS 156
ATOM      9  N    VAL   17   30.198  45.162  28.721  1.00 12.26              2TBS 157
ATOM     10  CA   VAL   17   30.956  45.745  29.835  1.00 11.31              2TBS 158

                    .                        .
                    .                        .
                    .                        .

ATOM   1655  CD2  TYR  245   14.549  34.212  -4.417  0.00 20.00              2TBS1803
ATOM   1656  CE1  TYR  245   11.808  33.996  -4.911  0.00 20.00              2TBS1804
ATOM   1657  CE2  TYR  245   14.087  33.477  -5.520  0.00 20.00              2TBS1805
ATOM   1658  CZ   TYR  245   12.715  33.368  -5.765  0.00 20.00              2TBS1806
ATOM   1659  OH   TYR  245   12.259  32.653  -6.831  0.00 20.00              2TBS1807
TER    1660       TYR  245                                                   2TBS1808
HETATM 1661  C1   BEN  246   31.488  48.310  20.312  1.00 20.30              2TBS1809
HETATM 1662  C2   BEN  246   32.812  48.214  19.951  1.00 19.75              2TBS1810
HETATM 1663  C3   BEN  246   33.111  47.223  18.971  1.00 22.63              2TBS1811
HETATM 1664  C4   BEN  246   32.216  46.350  18.252  1.00 20.35              2TBS1812
HETATM 1665  C5   BEN  246   30.913  46.593  18.786  1.00 14.49              2TBS1813
HETATM 1666  C6   BEN  246   30.513  47.518  19.788  1.00 25.62              2TBS1814
HETATM 1667  C7   BEN  246   31.077  49.169  21.257  1.00 19.02              2TBS1815
HETATM 1668  N1   BEN  246   29.800  49.347  21.673  1.00  8.66              2TBS1816
HETATM 1669  N2   BEN  246   31.893  49.980  21.938  1.00  9.08              2TBS1817
HETATM 1670 CA    CA   247   24.994  25.960  26.418  1.00 16.33              2TBS1818
HETATM 1671  O    HOH  301   26.953  44.930  36.915  1.00 43.47              2TBS1819
HETATM 1672  O    HOH  302   30.850  53.755  25.931  1.00 15.22              2TBS1820
HETATM 1673  O    HOH  303   26.187  50.990  24.323  1.00  5.60              2TBS1821
HETATM 1674  O    HOH  304   30.555  56.524  25.932  1.00 28.23              2TBS1822
HETATM 1675  O    HOH  305   33.910  53.136  24.036  1.00  9.45              2TBS1823
HETATM 1676  O    HOH  306   24.418  47.260  34.811  1.00 23.51              2TBS1824
HETATM 1677  O    HOH  307   23.257  61.631   9.074  1.00 50.82              2TBS1825
HETATM 1678  O    HOH  308   20.708  58.939  15.604  1.00 28.53              2TBS1826
HETATM 1679  O    HOH  309   17.357  31.529  24.895  1.00 24.87              2TBS1827
HETATM 1680  O    HOH  310   24.170  50.899  10.058  1.00  9.00              2TBS1828

                    .                        .
                    .                        .
                    .                        .

HETATM 1825  O    HOH  455   11.191  21.819  22.927  1.00 32.44              2TBS1973
HETATM 1826  O    HOH  456   16.546  63.660  20.602  1.00 33.22              2TBS1974
HETATM 1827  O    HOH  457   14.102  61.389  19.146  1.00 39.14              2TBS1975
HETATM 1828  O    HOH  458   32.772  34.783  25.141  1.00 30.84              2TBS1976
HETATM 1829  O    HOH  459   38.365  44.015  25.187  1.00 56.33              2TBS1977
HETATM 1830  O    HOH  460   36.179  39.288  21.439  1.00 41.13              2TBS1978
HETATM 1831  O    HOH  461   14.492  20.990   4.683  1.00 61.09              2TBS1979
HETATM 1832  O    HOH  462   25.637  19.443  16.555  1.00 64.61              2TBS1980
HETATM 1833  O    HOH  463   19.274  65.428  18.101  1.00 22.16              2TBS1981
HETATM 1834  O    HOH  464   19.470  32.777  24.901  1.00 17.10              2TBS1982
CONECT   50   49 1033                                                        2TBS1983
CONECT  193  192  307                                                        2TBS1984
CONECT  307  193  306                                                        2TBS1985
CONECT  399  398 1670                                                        2TBS1986
CONECT  439  438 1670                                                        2TBS1987
CONECT  457  456 1670                                                        2TBS1988
CONECT  848  847 1553                                                        2TBS1989
CONECT  890  889 1361                                                        2TBS1990
CONECT 1033   50 1032                                                        2TBS1991
CONECT 1118 1117 1224                                                        2TBS1992
```

```
CONECT 1224 1118 1223
CONECT 1299 1298 1463
CONECT 1361  890 1360
CONECT 1463 1299 1462
CONECT 1553  848 1552
CONECT 1661 1662 1666 1667
CONECT 1662 1661 1663
CONECT 1663 1662 1664
CONECT 1664 1663 1665
CONECT 1665 1664 1666
CONECT 1666 1661 1665
CONECT 1667 1661 1668 1669
CONECT 1668 1667
CONECT 1669 1667
CONECT 1670  399  439  457
END
.
```

```
2TBS1993
2TBS1994
2TBS1995
2TBS1996
2TBS1997
2TBS1998
2TBS1999
2TBS2000
2TBS2001
2TBS2002
2TBS2003
2TBS2004
2TBS2005
2TBS2006
2TBS2007
```

Index

New Compound Discovery with TRIPOS Products

Investigation of Individual Molecular Structures

The goal of molecular modeling was originally to build, visualize and compare 3D structures of new compounds or known lead structures with defined modifications. Therefore, Tripos software development started in 1979 with tools to sketch, display, rotate and manipulate molecules on the computer screen. Since that time, the sophistication of the software has increased and SYBYL now contains a broad range of tools for investigating the geometric, electronic and conformational structure of molecules.

Several levels for geometry optimization to cover a broad range of molecule classes are provided using molecular mechanics including the Tripos general force field, Amber version 4.0, MM2, MM3, or using semiempirical (for example AM1 and PM3) and *ab initio* quantum chemical methods via interfaces from Sybyl to programs such as MOPAC, ZINDO and Gaussian. Geometrical comparisons of two or more structures can be done by a rigid-body fit or flexible superposition (MULTIFIT).

Structures can also be retrieved from several frequently used structural databases such as the Brookhaven Protein Database (PDB), the Cambridge Crystallographic Database or MDL databases. Crystal structure data (fractional coordinates and unit cell dimensions) as well as different NMR raw data formats are readable and processable. If a database contains only 2D connectivity information, CONCORD rapidly generates 3D structures with one low-energy conformation per compound using a combination of knowledge-based and energy minimization approaches.

To further explore the conformational space that is occupied by a given molecule, different conformational searching techniques like Random Search, Grid Search and Systematic Search or a MOLECULAR DYNAMICS simulation (for example using Simulated Annealing with or without explicit solvation) can be applied.

Determining the charge distribution is an important step in modeling the intermolecular and intramolecular electrostatic interactions or in structure comparisons. Sybyl provides a range of methods including simple atomic charge estimation methods like the Gasteiger–Marsili approach and quantum chemical methods which can be used to compute the electronic structure of a set of molecules or may be chosen to calculate selected descriptors for molecular similarity or for structure–activity relationship investigations.

A wide range of properties such as electrostatics, hydrophobicity, hydrogen-bonding potentials and other local properties of a molecule and their subsequent fields can be calculated. Additionally, molecular volume and surfaces may be displayed and the various properties can be mapped onto molecular surfaces by color coding (MOLCAD).

As the field of molecular modeling has matured, there has been an increased demand for customization of methods, commands and menus to address the needs of individual researchers and project areas. Tripos has responded to this important request by providing a powerful Sybyl Programming Language (SPL) and by allowing to use customized force fields within the Force Field Engine (FFE). The FORCE FIELD ENGINE allows users to switch easily between different force fields (Tripos force field, Amber, MM2). More importantly, the FFE enables researchers to design and implement new or preferred specialized force fields without having to change and recompile any source code. Using FFE, force fields can be modified to include new energy terms and parameters. SPL is a flexible programming language that allows to implement new algorithms, create interfaces to external programs, and generally expand and customize the Sybyl environment to individual research needs. A large library of user-generated SPL applications and routines is available via the world-wide web. SPL is a major contributor to the open-system architecture of Sybyl and the high level of user flexibility available within this comprehensive and integrated software package.

Pharmacophore Identification in Absence of a 3D Protein Structure

Within pharmaceutical drug design, there is often little or no knowledge of the 3D structure of the target receptor. Therefore, chemists use the geometrical and electronic structure of compounds with known activity to intuitively define a working model of how the ligand interacts with its receptor or to develop simple structure–activity relationships based, for example, on a net charge required at a given location or on a required intramolecular atomic distance.

Several computational approaches are available to systematically compare molecular features and extract valuable information from a broad range of ligand data to describe aspects of the binding mechanism to the unknown active site of the target protein.

A first step in this direction might be to determine and display the common molecular volume of superimposed active compounds or the difference volume of inactive versus active compounds.

For a series of molecules that bind to a common active site, conformations can be identified which place the key atoms in the same or similar relative orientation. If these key atoms are obvious or known, the fast constrained conformational searching in RECEPTOR or the overlapping distance-based conformational space in the ACTIVE ANALOG APPROACH directly determines the bioactive conformations of each molecule in the data set.

When the key pharmacophoric centers are not obvious, the clique detection algorithm implemented in DISCO (DIStance COmparison) or the genetic algorithm provided with GASP (Genetic Algorithm Similarity Program) can generate both the aligned conformations of each compound and the models for the 3D pharmacophore – the spatial arrangement of key features shared by all active compounds. The GASP program represents an example for modern cross-platform access (for example Unix, IBM-PC, Macintosh) and for sharing information based on Web technology using the Internet and Intranets.

Another well-established approach to compare molecular properties and to determine their influence on the biological potency is QSAR (Quantitative Structure–Activity Relationships). The QSAR and the Molecular Spreadsheet functionality in Sybyl allow chemists to establish models using 2D or 3D structures. In classical 2D-QSAR analyses, either

(i) substituents or substructural units are assigned a constant contribution to the overall activity of a molecule (Free–Wilson, Fujita–Ban type) or (ii) physico-chemical (hydrophobic, steric, electrostatic) descriptors are used to find a correlation with the bioactivity (Hansch type). The presence of substructures can easily be detected by using the flexible SLN (Sybyl Line Notation) and a series of descriptor metrics for Hansch-type analyses can be computed by means of HDISQ (electro-topological state, Kier–Hall connectivity, E state, shape, symmetry indices *etc.*), CLOGP, CMR or Rekker's logP hydrophobic fragmental constant scheme or by using other property calculation techniques from the different Sybyl modules. The MOLECULAR SPREADSHEET unites the results of applications from different molecular research components and is central to all kinds of molecular analyses and is capable of extracting relevant information from the bulk of molecular data. It offers a unique framework for accessing, manipulating, combining and storing molecular property information tightly cross-linked with interactive graphs, visualization and computation capabilities. Also from the Molecular Spreadsheet, all the statistical analysis methods including the PLS (Partial Least Squares) method, factor analysis or clustering techniques are applicable to any data set.

Unlike 2D-QSAR, the CoMFA (Comparative Molecular Field Analysis) method directly takes the 3D molecular structures into account. Steric and electrostatic fields are calculated for the superimposed bioactive conformation of each molecule in the data set. The variance in these field data is used to explain the variance in the activity data by means of PLS and cross-validation statistics. The molecular regions that contribute most to the activity variations can be identified and give an indication of the pharmacophoric features. In addition, the predictive power of the QSAR correlations can be used to estimate the bioactivity of new compounds and to guide the modifications and refinements of existing structures to obtain a better activity profile.

The ADVANCED CoMFA module offers an automated routine for lead structure refinement as well as an interface to the fast SAMPLS program, new molecular field classes, or region focusing which enhances the signal-to-noise ratio in CoMFA models. Additionally, the GOLPE (Generating Optimal Linear PLS Estimators) program can be used to complement the CoMFA technology. GOLPE is a procedure based on an advanced chemometric method to establish optimal regression models for highly reliable predictions by using variable selection criteria.

As part of the rational drug design process, QSAR models are used to rank candidates for chemical synthesis and biological testing, to develop an understanding of the binding mode to proteins, as starting point for database mining, and to enhance the desired activity of a lead compound, once it has been identified.

3D Protein Structures, Ligand Binding and Biopolymer Modeling

The development of X-ray crystallographic and NMR techniques to determine an increasing number of protein structures has been important for the computerized drug design. Experimental 3D structures of proteins provide reliable information about the active site, the binding mode of ligands and can, therefore, serve as basis for a variety of receptor-based drug design approaches on the computer. Sybyl provides an environment for the general modeling and advanced design of biological molecules including proteins, peptides, nucleic acids, carbohydrates and lipids within the BIOPOLYMER module. In addition, a spreadsheet-based comparison and analysis of protein structures using well-established criteria as useful descriptors of local conformation, secondary structure and topological folds, solvent accessibility *etc.* can be performed with PRO-TABLE.

Protein structure determination can be accomplished using the TRIAD module within Sybyl for NMR data processing, bookkeeping, visualization, analysis of multi-dimensional spectral data and to extract structural information. CAPRI (Computer Assisted Peak Resonance Identification) quickly identifies spin coupling networks of proton NMR resonances and accurately assigns them to protons in the corresponding molecular structure. MARDIGRAS calculates a set of accurate distances from observed 2D NOE cross-peak intensities. DIANA (DIstance geometry Algorithm for Nmr Applications) determines conformations of biopolymers consistent with NMR measurements translated into distance and torsion constraints.

If only the sequence but no NMR or X-ray structure is available for a particular protein, homology modeling with COMPOSER generates a model structure based on the assumption that similar folding pattern exist for all members of a given protein family. Part of the tertiary folding pattern, the SCRs (Structurally Conserved Regions), is common to all of these homologs whereas loops as SVRs (Structurally Variable Regions) connect the SCRs. COMPOSER identifies the SCRs, determines the location of the SCRs in the target sequence, models the backbone of each SCR and constructs sidechains and SVRs by using established rules, protein homolog structures and database information. A different and complementary approach to predict a 3D protein model structure starting with its amino acid sequence is provided by the inverse protein folding algorithm in MATCHMAKER. A pseudo-energy function is used to describe the relationship between the amino acid sequence and the 3D protein structure and can be applied (i) to find all sequences compatible with a given 3D structure (all database sequences are threaded through a given fold) or (ii) by using a 3D structure database to find those 3D structures that fit a given sequence best.

Once a receptor structure is available, interactive DOCKING of a particular ligand under visual control with real-time energy feedback might be the first approach to get an idea about how well the ligand fits into the receptor cavity with respect to steric and electrostatic complementarity.

Alternatively, the *de novo* design program LEAPFROG builds ligand candidates that fit optimally into a rigid receptor cavity. First, potential interaction site points are created based on the 3D structure of the active site, then putative ligands or fragments are randomly retrieved from a fragment library and placed inside the active site to match site points and link existing ligand substructures. Movements of a given ligand candidate within the cavity are taken into account. The quality of the ligand-to-site fit is evaluated based on the computed binding energy. Portions of the ligand with non-optimal fit are modified and refined until an optimal binding is found with respect to steric, electrostatic and hydrophobic contributions. Considering the variety of possible fragments and their combinations in the iterative refinement process, LEAPFROG is able to propose new lead structures and to improve initial leads.

Database Mining and Lead Compound Discovery

To identify a new lead structure, a broad selection of compounds from different molecule classes is typically tested. High-throughput screening (HTS) has dramatically increased the number of compounds that can be screened in a given time. Considering the large number of available and potential compounds and the limited resources of every company, the need of useful and efficient compound selection techniques is evident. Depending on the available information, there are three general ways in which computers can assist in selecting compounds from one or more databases or libraries: pharmacophore searching, receptor-based searching and molecular diversity selection.

The 3D pharmacophore, as the structural prerequisite for each active compound, is often represented by individual functional groups and their relative orientation to each other encoded in a set of interatomic distance constraints. Once determined, a 3D pharmacophore model can be used for a UNITY substructure search through an electronic database of 3D structures. During this *pharmacophore search*, the conformational flexibility of the structures in the database is taken into account using the fast directed tweak method (minimization in torsional space). This allows UNITY to identify all new lead candidate structures which are able to adopt the pharmacophore geometry, independent of whether the bioactive conformation is stored in the 3D database or not. Beyond flexible 3D substructure searching, UNITY allows fast 2D substructure and exact-match searching, non-structural data searching in relational databases (biological and physico-chemical data), similarity searching and offers very flexible query definitions (Markush atom definitions including nesting).

A known 3D receptor structure can be used to search a database for molecules which might bind into the receptor cavity. The steric, hydrophobic and electronic complementarity can be investigated by computing the binding energy of all potential ligands. Using UNITY integrated in Sybyl for *receptor-based searching*, a query can be created where active-site locations occupied by receptor atoms and, therefore, not accessible to ligand atoms, are taken into account as excluded volume. This query definition is then used by UNITY to search for molecules which could fit into the receptor pocket. Any pharmacophore information can also be included in this receptor-based database mining.

In the absence of any 3D structural information about the pharmacophore or the receptor binding site, a careful selection of compounds to be screened should cover a broad range of structurally diverse molecules and, thus, increase the screening hit rate. To assist in high-throughput chemistry and screening, Tripos provides a software package called the Molecular Diversity Manager (MDM) for the design of compound libraries with optimal diversity and *molecular diversity selections*. MDM consists of several components: LEGION is used to create and store virtual combinatorial libraries based on reactants or on a core product molecule. LEGION handles possible complexities such as multiple attachment points or nested and merged substitutions. SELECTOR is used to calculate appropriate descriptor metrics and subsequently select a diverse compound subset. A variety of metrics such as 2D fingerprints, logP, HDISQ, molecular fields and pharmacophoric triplets can be computed and then used in compound library selection, refinement and optimization. DIVERSE SOLUTIONS provides distance-based algorithms and a novel cell-based selection technique including the novel BCUT descriptor values. The cell-based approach allows fast selections from large databases comprising hundreds of thousands of structures. It will also identify under-represented or empty diversity regions (voids) in a given compound library and can assist in filling these diversity voids with compounds from other existing databases.

A screening library with broad molecular diversity and undefined structural objectives for initial lead structure generation is available with the OPTIVERSE COMPOUND LIBRARY. This library was designed by applying Tripos' proprietary technology, synthesized in cooperation with Panlabs, Inc., and contains nearly one hundred thousand of chemical substances in 96-well microtiter plates with a single compound per well. In addition to molecular diversity, selections of reactants and products are based on stability, chemical reactivity, reaction yield, toxicity, hydrophobicity, molecular weight and other criteria. Every new compound adds different structural information, avoiding redundancy and holes in the molecular diversity space.

This library also forms one basis of the new-compound discovery service that Tripos offers based on its pre-eminent position and broad expertise in the application of computer technology to molecular discovery. The value-added research service capitalizes on

(i) the SYBYL package providing an integrated science, analysis and chemical information software for modern network architectures supporting a distributed computing and graphics environment, and

(ii) TRIPOS' experience in using these software tools, for example, in rational drug design, molecular information analysis, proprietary design strategies for both lead-discovery and lead-refinement libraries.

TRIPOS GmbH, Martin Kollar-Str. 15, D-81829 Munich, Germany
phone: ++49-89-4510300, fax: ++49-89-45103030
TRIPOS Inc., 1699 South Hanley Road, St. Louis, MO 63144, USA
phone: US-800-323-2960, fax: US-314-647-9241
http://www.tripos.com